A LOUISIANA COASTAL ATLAS

THE NATURAL WORLD OF THE GULF SOUTH

Craig E. Colten, *Series Editor*

A LOUISIANA COASTAL ATLAS

Resources, Economies, and Demographics

Scott A. Hemmerling

LOUISIANA STATE UNIVERSITY PRESS
BATON ROUGE

THE WATER INSTITUTE
OF THE GULF

Published with the assistance of the Water Institute of the Gulf.

Published by Louisiana State University Press

Manufactured in China
First printing

Designer: *Mandy McDonald Scallan*
Typefaces: *Haboro Slab, text; Gotham, display*
Printer and binder: *Shenzhen Caimei Printing through Four Colour Print Group, Louisville, Kentucky*
All maps were created by the author.

Library of Congress Cataloging-in-Publication Data are available at the Library of Congress.

ISBN 978-0-8071-6588-1 (cloth: alk. paper)

The paper in this book meets the guidelines for permanence and durability of the Committee on Production Guidelines for Book Longevity of the Council on Library Resources. ♾

I dedicate this work to my daughter Phoebe.

Watching her grow up has been a constant reminder of why I do the work that I do. I truly believe it is our duty to make sure that we leave some positive and lasting impact on the world. If this work inspires even one child, then I have accomplished my goal.

CONTENTS

ACKNOWLEDGMENTS

Seeing this project through from inception to completion was made possible by a number of individuals. First and foremost, I am greatly indebted to Craig Colten, whose vision of a mini-atlas of coastal change was the seed that grew and transformed into this work. We quickly realized that the stories I wanted to tell cartographically were much more expansive than we had imagined. Craig's continued guidance proved invaluable, providing much-needed focus and always reminding me to see the forest for the trees.

Many thanks must also go to Bethany Garfield, who was essential in the data gathering and preparation phase of this research. Bethany spent many hours scouring historical data sources and transcribing these data. Truly, Bethany singlehandedly did the work of an entire team.

Special thanks go to the staffs at numerous public libraries across the state. These individuals are often the unsung heroes of historical works such as this one. Additional data were provided by Brady Couvillion, of the U.S. Geological Survey; Audrey Maass, with the LSU Department of Geography and Anthropology; and Hal Needham, formerly with the Southern Climate Impacts Planning Program (SCIPP) at LSU. Special thanks also go to staff at the NOAA Southeast Fisheries Science Center for their assistance in locating many of the historical reports needed for this atlas.

Many individuals provided insight as this project took shape. Denise Reed and Ehab Meselhe of the Water Institute of the Gulf gave of their time and offered valuable critiques on the initial draft. Their scientific insights only served to strengthen the final product. Phillip LeFargue, also of the Water Institute, gave expert guidance on the overall look and style of this atlas.

Generous support for this project came from the Baton Rouge Area Foundation (BRAF) and the Louisiana Coastal Protection and Restoration Authority (CPRA). I am extremely grateful for the insights provided by Karim Belhadjali, Andrea Galinski, Mandy Green, Melanie Saucier, and several other individuals with the CPRA. Their ideas on how the data being gathered as part of this study could be used by coastal planners and scientists gave essential grounding to this book.

I acknowledge with much appreciation the contributions of LSU Press. The support of Margaret Lovecraft and Alisa Plant has proven invaluable in seeing this project through to completion. Early comments from the LSU Press production staff as well as the anonymous reviewer helped shape the overall look and feel of this book. They took the vision in my mind and helped me get it on the page.

Finally, I would be remiss if I did not acknowledge my wife Nicole and my daughter Phoebe. This book would not have been possible if not for them. Completing this work often required long hours spent away from friends and family. At the same time, friends and family provided me with the inspiration to keep moving forward.

A LOUISIANA COASTAL ATLAS

INTRODUCTION

Craig E. Colten

LOUISIANA'S COASTAL RESIDENTS are as dynamic as the wetlands where they live. Over the millennia, the Mississippi River built a series of deltas, and each began receding after the river abandoned the older course in favor of a new active channel. As a result, there has been no fixed coastline during the past 10,000 years. Instead, the shore has been constantly shifting. Prehistoric populations followed the advancing shoreline in quest of coastal foodstuffs. The arrival of Europeans and Africans in the eighteenth century dislodged many of the indigenous people and initiated a major human resettlement of the region. Devastated by disease or violence, remaining Native Americans often migrated to new locations. Meanwhile, French, African, German, Spanish, and English settlers built farms and cities as close to the coast as they could find firm footing. Over time, some settlements disappeared as others thrived. While each group of settlers established deep attachments to their Louisiana home, none has been firmly fixed in place. Population patterns over the past sixty years illustrate the ever-changing nature of our coastal settlements (fig. 1).

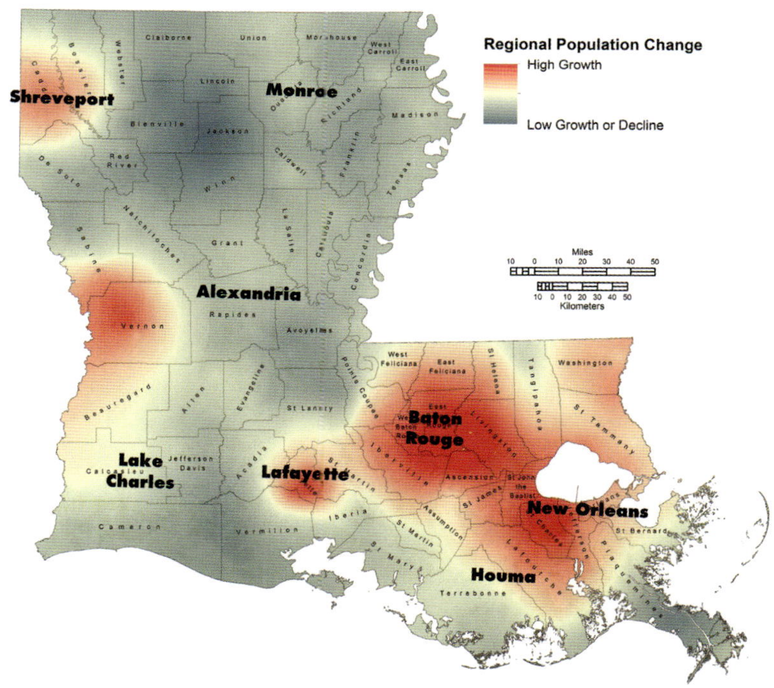

Fig. 1. Regional Clusters of Community Growth (1950–2010)

Cluster analysis of population change in census-designated places between decadal censuses (1950–2010) reveals significant clustering (z-scores higher than 2.58 standard deviations).

Data Source: U.S. Census Bureau

Local livelihoods and economies have reflected ecological, cultural, political, and technological shifts. Cypress harvests have followed an erratic history driven by changing technologies, demand, and, ultimately, by the near elimination of the resource. Colonial experiments with indigo and tobacco gave way to successful cultivation of rice and sugar along the rivers and bayous, resulting in changing economies of agriculture. Technological changes in cultivation and processing affected crop production in the coastal parishes and also decreased the need for on-site labor and hence altered the region's population patterns. Shrimping, which was a negligible activity until the late nineteenth century, now is one of the prime fisheries and has propelled population growth in fishing ports. Oyster harvests rose with conservation policies in the early twentieth century and contributed to fishing and processing settlements on the margins of *terra firma*. Tropical cyclones and human-made disruptions have prompted the demise or retreat of some fishing communities and coastal resorts. The rise and fall of each resource-based economic pursuit has impacted local communities and prompted adaptations. From a long-term perspective, the population and economic geography of the coastal region has never been static.

Oil extraction is a relatively recent enterprise and has undergone a series of dramatic fluctuations. From onshore activity near Crowley early in the twentieth century, the industry expanded into the nearshore waters of the coastal parishes by the 1930s. By the early 1970s, although many of the accessible onshore deposits had been exhausted, offshore oil enabled a new series of booms, followed by the bust of the 1980s, and then renewed activity in the Outer Continental Shelf (OCS) by the 1990s. Changing global supplies and prices continue to impact local drilling activities in complex ways. Population adjustments and economic adaptations have followed each successive petroleum boom-and-bust.

Demographic and economic episodes may be briefer than geologic epochs, but they are every bit as dynamic as land loss and land building. Louisiana has experienced innumerable episodes of coastal population and economic expansion and contraction. Residents have advanced toward the coast to harvest wetland, estuarine, and marine resources, and they have also relocated landward in the face of coastal hazards or resource depletion. These oscillations demonstrate that mobility and adaptation are two of the most deeply entrenched traditions of the coastal population.

The current environmental situation that has given rise to coastal restoration programs is not limited to the ecological, geological, and hydrological processes that produce land loss. Humans make the decisions, they design the projects, and they cope with the consequences. Indeed, the combined processes of coastal land loss and coastal restoration will create conditions impelling residents to adapt once more. Coastal restoration projects that seek to preserve valuable natural resources can and should encompass cultural restoration as well. An effective and acceptable coastal restoration must consider the human impacts alongside environmental consequences. This atlas foregrounds human populations as a dynamic part of coastal Louisiana. While not omitted from planning efforts, the human dimension often has been secondary to important and complex scientific and engineering considerations. Yet, science and engineering can be more effective in reshaping our coastal environment when they incorporate the equally important and complex human dimension.

Previous adaptations have been in response to natural processes, such as tropical cyclones, or the indirect outcomes of human-driven processes, such as levee building or global economic cycles. In the course of historical disruptions, there has been little government assistance with adaptations. In the case of coastal restoration, there is an opportunity to foresee human impacts and to take steps to mitigate them or to aid with adaptations. Whenever we discuss coastal restoration, we must also discuss coastal culture restoration.

RESOURCE USE AND ADAPTATION

CYPRESS

Grand cypress stands covered extensive bottomland swamps when the French arrived to occupy the lower Mississippi River valley in the 1710s. Forest clearance was essential to open space for agriculture and settlements and to secure building materials. Intending to establish wealth-producing plantations, colonists felled cypress stands and used the easily worked wood to build homes and other structures. Early agricultural pursuits yielded little of the fortunes envisioned by Louisiana's early investors, but settlers discovered a market for the exceptionally durable cypress. By the 1720s, the colony was producing planks and shingles for export to the West Indies, thus establishing one of the colony's earliest economic foundations. Related enterprises emerged to support this extractive activity. Sawmills tapped wind or horse power until a sufficient levee system was erected to protect floodplain farmland. During high river stages, water could be released through flues in the levee to turn water wheels and power saws. Thus, flood-protection structures complemented the emergent timber industry and accelerated cypress removal, but the activity remained concentrated in the forests near the river and the mills.

Gradually, cypress harvests pressed farther into the back swamps surrounding New Orleans to obtain timber for export to the Caribbean. Although floods and international political events could temporarily disrupt local harvests and external demand, cypress remained an economic cornerstone of the colonial economy and accounted for approximately 25 percent of export values in 1763. Managed by French landowners, enslaved Africans, as well as escaped former slaves living in the wetlands, did most of the difficult labor

(Usner 1979). By the end of the French period, the impact of their efforts was obvious to many visitors, who described the denuded wetlands in the vicinity of New Orleans.

When Spain took over the colony in 1763, it confined exports to its territories. This meant a loss of former French colonies as customers and contracted the market, threatening the timber industry. Loss of revenue, however, prompted Spanish officials to resume trade with the French West Indies, which reinvigorated the cypress business. As sugar emerged as a viable crop late in the eighteenth century, demand for cypress wood to make "sugar boxes" arose and provided an important local market. By 1800, cypress lumber produced in the American territory of Florida was severely undercutting prices of the product taken from Louisiana. This international competition diminished the demand for a commodity that had been an economic mainstay. Adjustments in foreign demand coupled with local apprehension of wood construction following a pair of devastating fires in New Orleans in 1788 and 1794 further eroded the viability of the business. The very resource that sustained the colony's economy during its early years dramatically declined in importance on the eve of the Louisiana Purchase and left many who had relied on it in precarious financial situations (Moore 1983a). Yet, most timber was taken from private lands that also produced agricultural commodities, and the mills were relatively small operations. Vagaries in the global market did not cause an economic collapse or drive major migrations.

Cypress harvests reaccelerated after the United States acquired the Louisiana Territory in 1803. Political realignment created new opportunities for cypress impresarios as new commodities fueled economic growth. Prosperity among sugar and cotton planters produced demand for timber and also sparked impressive growth in New Orleans, and the attendant use of cypress lumber. Even though urban residents were apprehensive about the fire risk linked to wood construction, cypress remained an important component in commercial and public buildings and also was widely used in domestic structures (Moore 1983b).

Emerging demand drove the timber harvest to areas previously lightly worked. Steam-powered sawmills became more common by the 1840s and detached processing from the banks of the Mississippi. This technological innovation enabled further expansion of cypress exploitation. Yet, during the early years of the American period, taking cypress from public lands—where most of the unworked stands were located—was illegal before the completion of federal land surveys. Demand nonetheless drove illicit harvests that ineffectual federal enforcement was unable to stem. In 1849, with the passage of the first Swamp Lands Acts, the U.S. Congress granted millions of wetland acres to states, and they quickly went to work disposing of their swamplands to private owners (Eisterhold 1972 and Moore 1983b). Thus, by the mid-nineteenth century, cypress extraction was pressing farther into Louisiana's bottomland forests in the Atchafalaya and Maurepas basins.

The industrial-scale heyday of the cypress industry occurred during the late nineteenth century. A combination of capital transfer from the upper Midwest timber industry and more efficient technologies for tree removal enabled a true boom period from about 1895 to 1925. Northern timber companies purchased sizable tracts of land in the Atchafalaya and Maurepas basins. They introduced pullboats and built narrow-gauge railroads to move the giant trees to mills on the wetland margins. A harvest of 45 million board feet in 1879 grew to 500 million in 1894 and crested at over 1 billion board feet in 1913 (Mancil 1969, 23). Mill towns at Patterson, Ponchatoula, Lutcher, Harvey, Berwick, and Galliano thrived as woodsmen cleared the surrounding swamps. The aggressive harvests left a barren landscape. By 1934, the U.S. Forest Service reported that of the 1.7 million acres of cypress lands in Louisiana, more than 1.6 million had been denuded (Norgress 1947, 69–71). As the supply of cypress disappeared, mills shut down and timber-dependent communities lost population and economic viability. There were 150 cypress mills operating in 1915. By 1920, that number fell to ninety, and it continued downward to four in 1940 (fig. 2). None survived in 1956. Workers had to turn to different pursuits as sawmills fell into disrepair and their blades gathered rust. Smaller sawmill towns, like Plaquemine and Patterson, experienced more serious declines than larger communities with a more diverse economic base. Seven of the fourteen towns founded as sawmill communities became ghost towns after the resource was exhausted (Mancil 1972). In the absence of effective conservation practices for the slow-growing cypress, this was an industry that was never to return. As the mill towns dried up, former workers moved to other locations to pursue other livelihoods.

SHRIMP

Shrimping is unquestionably a major economic pursuit in Louisiana's coastal parishes, but it has not always held such a prominent position. Early shrimping was labor intensive as fishermen used seines in shallow estuaries. This activity involved a limited number of fishermen who operated seasonally and worked out of small ports near the bays and estuaries. There were no dominant

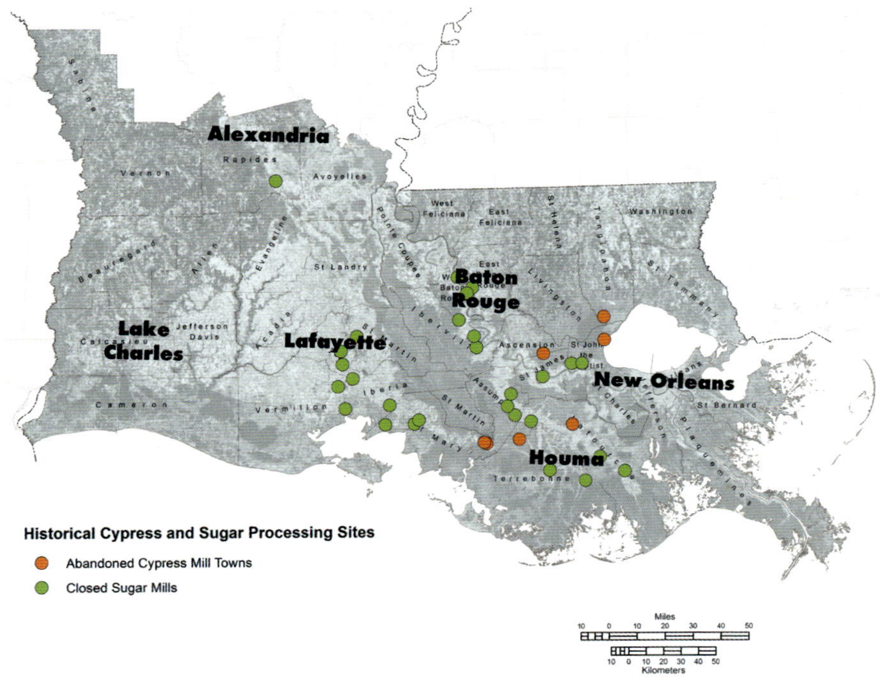

Fig. 2. Abandoned Cypress Mill Towns and Sugar Mills
Data Sources: Mancil 1972; Rehder 1999

shrimping centers. A growing national demand, along with new fishing and processing technologies, has accelerated shrimping since the 1930s. Technology was a key influence in expanding shrimping activity. Small gasoline-powered trawls enabled more effective shrimping beginning in 1917, and the state's catch jumped fivefold between 1919 and 1937. A proliferation of small family operations accompanied this expansive phase and drew in fisherfolk from a range of ethnic groups—Acadians, Anglos, Native Americans, Isleños, African Americans, Filipinos, Chinese, Italians, and Malayas. Much of the early catch was dried for export and thus gave rise to isolated, export-oriented processing settlements inhabited by Filipino workers, such as Manila Village. Small communities benefited as fisherfolk and their families purchased boats, gear, fuel, and other necessities. Thus, during the interwar years, coastal communities experienced unprecedented economic benefits from the shrimp-fishing industry (Becnel 1962, Keithy 1991, and Davis 2010).

With the discovery and exploitation of nearshore shrimp grounds in the Gulf during the late 1930s, small ports saw accelerated growth. Morgan City's position near these offshore shrimp resources contributed to the expansion of its role as a center for processing and distribution (Marks 2012). Houma, with a different situation at the apex of a system of bayous linking it to marine resources, also arose as an important shrimp processing and distribution center (Padgett 1969).

In the post-World War II years, consumer demand for shrimp increased. Market demands and the pursuit of large shrimp offshore prompted swift adaptation by shrimpers. Investing in deepwater vessels that were able to remain away from port for several weeks, fishing families began harvesting the offshore water. Use of ice, which began in the 1930s, enabled extended fishing trips. New fishing practices were distinct from the nearshore fishing activity and stimulated shipyards and associated operations offering gear and supplies. Onshore support activities increased their scale of operations and became more centralized. Processors shifted from dried, salted shrimp to frozen product, causing the collapse of Manila Village after Hurricane Betsy in 1965. Communities such as Houma and Morgan City, however, benefited from the consolidation and increasing scale of fisheries.

Growth in shrimping did not continue on an uninterrupted trajectory. Inconsistent prices in the 1950s disrupted the fishery's development, and each price decline rippled through the shipyards and other support industries. Shrimp landings fell dramatically in the mid-1950s and early 1960s due to poor prices, causing household incomes to suffer and unleashing a round of boat repossessions by banks. Boom-and-busts continued through the 1970s (Marks 2012). Diversion of freshwater through the Morganza Spillway in 1973 and oil spills in the late 1970s and mid-1980s introduced complications to the fishing communities, who faced a potential loss of business due to fear of either reduced populations or tainted catches (Colten et al. 2012). Shrimping was never a steady trade, and local economies fluctuated in accordance with the success of the fishing families.

The global economy became a powerful force in shrimping in the last two decades of the twentieth century. The oil bust of the 1980s prompted many former oil-field workers to turn to shrimping as high prices during the 1970s appeared to promise a lucrative income. Joining the influx of local residents, Vietnamese refugees arrived on the scene in the mid-1970s. Those who had been shrimpers in their homeland sought similar opportunities in the United States and added to the number of boats in the fleet. Fear of overharvesting by the burgeoning number of fisherfolk prompted passage of the Fishery Conservation and Management Act, which led to the formation of a fishery management council for the Gulf of Mexico. Unfortunately, the council took few actions to limit the harvest (Marks 2012). Fisherfolk and Louisiana politicians resisted efforts to impose restrictions on fishing gear to protect endangered sea turtles. When regulations on fishing gear finally took effect in the 1990s, the costs drove many shrimpers who had entered the industry during the rapid expansion of the 1970s out of business. Some who had maintained longstanding family businesses adapted to the new requirements and were able to continue. Nonetheless, the number of industry fishermen in the late twentieth century continued to shrink. As part of that retrenchment, the number of onshore processing plants declined from 48 in 1975 to 31 in 1997 (Marks 2012).

In the late 1990s, shrimping experienced a brief spike. Falling diesel fuel prices along with a rapid rise in shrimp prices encouraged many shrimpers, especially younger captains and Vietnamese families, to enter the field. Many took out loans to purchase large trawlers. After 2001, however, fuel prices took an upward turn and a simultaneous drop in shrimp prices strained the finances of newly debt-burdened fisherfolk. With the massive disruption caused by the hurricanes in 2005 and 2008, along with the BP oil release in 2010, catches declined, further undermining shrimping's profitability in Louisiana waters. Added to the domestic challenges, the influx of cheap foreign shrimp dramatically undercut the viability of Louisiana shrimpers (Marks 2012). Even the massive public relations campaign to restore the image of Gulf shrimp after the 2010 oil release could not restore the cost competitiveness of domestic fisherfolk. For many years, shrimpers had survived through the sweat equity of family members working as crew. Yet, in the long run many shrimpers and their children are abandoning the trade and seeking other forms of employment (Colten et al. 2015). Declining demand for boats, nets, other gear, fuel, and ice ripples through shrimp ports, and falling catches impact processing centers. There has been no long-term stability in the shrimp industry, and its future, while Americans delight in consuming these creatures, is not all that promising as a family enterprise on the Louisiana Gulf Coast.

SUGAR

Sugarcane became the most important commercial crop along the lower Mississippi River and its distributaries, notably Bayous Lafourche and Teche, during the early nineteenth century. Acreage in other crops dominated the lower river floodplains, but experimentation with sugarcane began at least by the 1740s. Several factors were necessary for sugar to gain a prominent role in the colony's agriculture. Sugarcane requires well-drained soil, and this necessitated adequate levees to fend off seasonal high water. Limited success with high-value export commodities prompted farmers to consider other options, and the collapse of indigo due to insect infestation in the 1790s increased interest in a substitute crop. Immigration during the Spanish period boosted the colony's population, both free and enslaved, and this human component provided the essential labor force. Outside the colony, there was ample demand for sugar, and with the development of a profitable granulation process in 1795, the industry took root. Sugarcane largely replaced the experiments with indigo, tobacco, and other crops as it spread up- and downriver from New Orleans and reached Pointe Coupée Parish before Louisiana statehood. In addition, the revolution on Saint-Domingue (Hispaniola) led to the immigration of experienced planters and slaves to Louisiana. The disruption on the Caribbean island nearly halted production there and enabled competing sugar colonies to thrive, thus opening opportunities for Louisiana planters. Following the arrival of exiles from Saint-Domingue, Louisiana plantations exported up to 5 million pounds of raw sugar a year by 1801 (Vaughan 2003, 95). Anglo planters from the Eastern Seaboard spread the crop to the banks of Bayou Teche and into Terrebonne Parish by 1820 (Rehder 1999, 41-48). Along the Mississippi, Anglo planters bought out the *petit habitant* Acadians who had acquired land during the Spanish regime. As planters cultivated more acres, production rose from 69,000 tons annually to over 110,000 tons in 1860 (U.S. Census). Not only was there a transformation of the agricultural landscape, but the social landscape also changed dramatically. New settlers displaced the existing Acadians, many of whom moved west into the Atchafalaya Basin (Estaville 1993). Meanwhile, the number of African Americans swelled on the labor-intensive plantations.

Production suffered during Reconstruction, but planters upped output to unprecedented levels—3 million tons a year—by the turn of the century. By 1982, production had climbed to over 7 million tons. Labor during the twentieth century came from the descendants of former enslaved workers and also from Italians recruited to work the fields. Tariff protection provided a vital incentive to planters. With economic safeguards, planters were able to transition to a wage-labor situation. Statewide acreage planted varied and reached an early peak of 345,000 acres in 1964. Acreage fell considerably during the 1980s, but rebounded and attained new heights of over 490,000 acres by 2000. Nonetheless, the number of sugarcane farms declined dramatically between 1971 and 2012. There were over 1,500 farms in the early 1970s and only 475 by 2012. This change reflects increased mechanization and the ability of one farm operator to cultivate more land. Demand for labor has declined, as have the number of sugar mills. From a peak of 1,536 in 1849, the number of sugar mills dropped to 300 in 1900 and eroded to only 13 by 2005 (fig. 2) (Rehder 1999, 129). Economies of scale and improved transportation equipment mean that fewer larger mills can process more cane than was possible in the nineteenth century. With the closure of grinding mills, many resident laborers moved from farms to nearby towns or to larger cities. Rural populations declined as mobile laborers adapted to changing economic opportunities.

Despite elimination of slavery and loss of jobs on sugar plantations, African American communities remain. Yet with the arrival of petrochemical works to the Mississippi River floodplain, a new impetus for people to leave rural communities has emerged. After acquiring prime riverfront property, some large manufacturers subsequently purchased and relocated entire communities to avoid the risks associated with accidental releases of toxic fumes or explosions and fires. Morrisonville near Plaquemine is a well-known example of an established African American town that was uprooted to make way for Dow Chemical's plant. Some "fencerow communities" that did not receive assistance have struggled to secure relocation funds to escape the risks and discomforts that accompany proximity to major petrochemical plants (Colten 2012). Thus the expanding footprint of the petrochemical complex has dislocated established rural communities and prompted another adaptation to changing economies.

Sugar no longer provides the number of jobs that it once did, and the landscape along the state's rivers and bayous now exhibits a juxtaposition of crops and petrochemical works. Levees once built to protect sugarcane from inundation now provide flood security for major industrial operations set amidst lush cane fields.

OIL

Agriculture and forestry have had profound impacts on the landscapes of Louisiana's coastal zone parishes. Both activities are land extensive and economically significant. Despite their prominence, however, a newly recognized resource gave rise to a dramatic series of social, economic, and biophysical impacts across the state in the early twentieth century. Oil discoveries near Jennings in 1901 and near Shreveport in 1904 unleashed aggressive searches to find and tap subsurface petroleum deposits. Initially, extraction focused on deposits found on the flanks of onshore salt domes, particularly north of the coastal wetlands where drillers found stable geologic foundations for the heavy oil-field apparatus. Jennings and New Iberia were among the early major finds in southern Louisiana. Exploration and successful drilling expanded the producing area into the entire coastal zone by the 1930s, from Plaquemines to Calcasieu parishes. Realizing the potential revenue from this mineral resource, the state imposed a severance tax on oil production in 1928 and consequently tied its economic well-being to this mineral.

The emergence of oil extraction in the wetlands in the 1930s recalibrated the economic value of the coastal marshes. Land used for trapping had lost value during the Great Depression, but the presence of oil boosted the prospects of large landholding companies, which suddenly became much more valuable. The industry's arrival spawned a series of temporary camps where workers lived between shifts on the land-based rigs (Davis 2010). The economic impacts radiated outward from the rural fields, and towns like New Iberia, Lafayette, and Morgan City experienced considerable growth as they attracted outside investment in industries such as drilling equipment, exploration, and related services. Offshore drilling began in the 1930s and expanded dramatically in the 1960s and 1970s. Terrebonne Parish's population, for example, tripled during the period from 1930 to 1980, reflecting the petro-boom's influence. Onshore production peaked in the 1970s, and there was fear that the economic flower had lost its bloom. Expansion of offshore production, however, sustained the regional economy until the 1980s and drove urban growth in the coastal region. Oil renewed population increases in Morgan City and more than replaced the defunct cypress lumber industry in local economic importance. As the favored site for rig fabrication, shipbuilding, and related service industries, Morgan City became a prominent hub for the offshore industry. New Iberia and Houma, with dredged channels connecting them to offshore activities, also grew during the boom years, as did Lafayette, which hosted exploration and other industry services that could operate at a distance from the actual resource. These communities saw energy-related employment rise in importance and truly shape their economies. Employment in mining in the coastal parishes crested in 1980 (Pulsipher 2006).

Oil permeated every aspect of life in the coastal parishes. During the early years, trappers and fishermen guided oil exploration teams into the wetlands. Oystermen contracted with pipeline companies to develop ways to minimize the impact of pipelines on oyster leases (Theriot 2014). Shrimpers delivered supplies to offshore rigs. As the state experienced net in-migration during the

1969–1983 period, restaurants, realtors, and other businesses benefited from the influx of out-of-state companies and their workers. Shipbuilding firms manufactured specialized craft for the industry. Pilots, divers, and other specialists found a demand for their skills offshore, to say nothing of the countless roustabouts who worked on the rigs during the 1970s. Few industries offered such rewarding salaries to those without a high school diploma.

During the peak years, population and industrial growth placed heavy demands on communities in terms of housing, public utilities, medical services, and public safety. During downturns, the same communities struggled to sustain the level of services they put in place during boom years (Coelho 2006). In the global oil glut of the early 1980s, coastal Louisiana suffered a dramatic bust. The Houma metropolitan area lost nearly a quarter of all jobs from 1982 to 1987. With the drop in oil-related employment, there was a net migration out of the region that continued into the 1990s (McGuire 2006, 31). The industry consolidated many of its office-bound activities, like geotechnical analysis, in Houston, Texas, and many higher-paying technical positions left the state. Service industries also went through painful contractions (Pulsipher 2006). State revenue suffered as severance tax income declined, although increased oil prices offset the tax revenue downturn to a degree. Personal incomes flattened out during the 1980s as the region searched for alternatives to oil.

Expansion into the OCS in the 1990s renewed the importance of oil to the region's economy, but also prompted adjustments. Venice, which was a key jumping-off point for nearshore rigs in the 1970s, has lost most of its oil-field service capacity to Port Fourchon—a made-to-order service base with no permanent population. The fixed platforms used in the 1960s and 1970s gave way to new deepwater drilling technologies and fewer drilling platforms; some fabrication shifted to out-of-state locations. Morgan City now hosts a much-diminished rig-building industry. Furthermore, OCS drilling took place beyond the state's territorial waters and thus did not pay severance tax. So despite this activity's expansion, the state did not experience a comparable influx of revenue. Post-Katrina legislation will initiate some revenue sharing from the OCS to the state in 2017. During the most recent boom, jobs in the industry increased and local services enjoyed renewed activity. In Terrebonne Parish, 1997 employment in almost all oil-related employment sectors was higher than in 1980 just before the oil bust (McGuire 2006, 32–33). Oil prices began to climb in the early twenty-first century and stimulated additional growth of OCS activity, with spillover economic effects in Louisiana's coastal parishes. After several strong years, the BP oil gusher disrupted mineral-related activity briefly in 2010, but the industry remains a prominent part of the economy, even though the impacts to onshore life have certainly changed since the heyday of the 1970s. Also, hydraulic fracturing has opened new inshore marine shale oil resources, which has fueled great expectations for a new oil boom across the state. Falling oil prices in 2014, however, raised questions about the immediate future of the "fracking" boom.

HUMAN ADAPTATIONS IN A PERILOUS PLACE

Adaptation and mobility have been ingrained survival strategies of residents along the northern Gulf of Mexico coast for millennia. As the delta built up soil in the Gulf during prehistoric times, Native Americans followed the formation of new land southward. French, Spanish, African, English, and other European settlers arrived in the eighteenth century and launched a series of economic experiments as they adapted to a new environmental setting. Most trial crops, like indigo and tobacco, offered short-lived benefits. Other commodities, such as sugar and rice, had greater staying power, but were not stable economic drivers given climatic and market vagaries. External political and economic pressures have prompted both human and economic mobility and adaptation. Natural resource exploitation experienced periods of rapid expansion, often followed by collapses due to depletion or external forces. Cypress lumbering denuded wetland forests, oil has seen fluctuations that have taken coastal residents on a roller coaster ride, and shrimping has endured similar periods of expansion and contraction. Each economic pursuit has prompted changes in the settlement landscape of the coastal region. Periods of prosperity have stimulated population growth, mostly in urban areas, over the last century. Hazard events, price collapses, and resource depletion have forced people to abandon coastal residences. All the while, new pursuits have sustained an adaptable population that has been willing to relocate when the need arose. Viewing the coastal region as one that is highly dynamic is essential to prepare for its future.

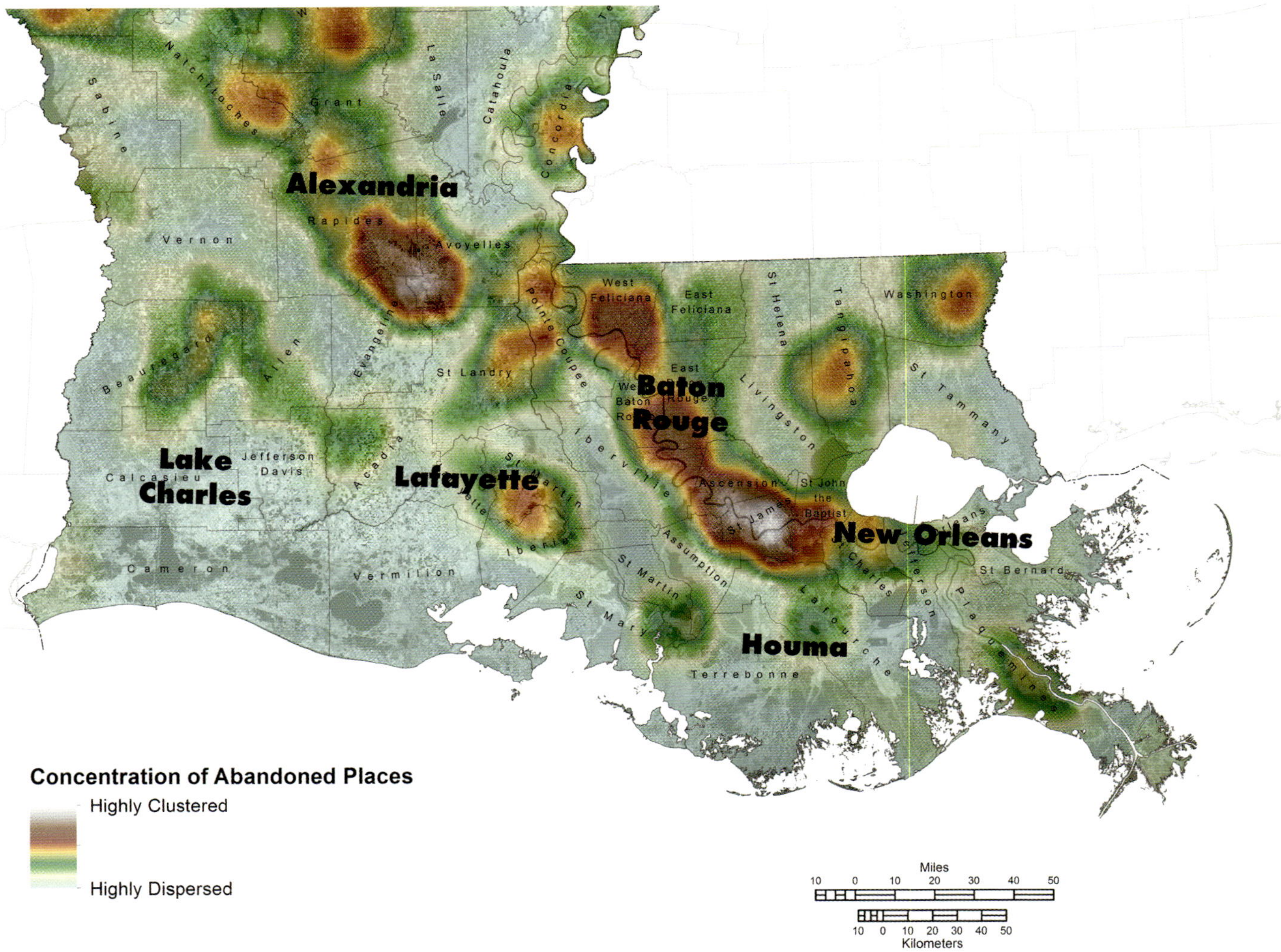

Fig. 3. Abandoned Towns and Communities

Analysis of the U.S. Geological Survey (USGS) inventory of place names removed from the agency's maps indicates a concentration of deleted place names in areas with forestry and agricultural economies. Prepared using a cluster analysis of USGS data.

Towns suffered population losses for a variety of reasons. Chenier Caminada was abandoned after the hurricane of 1893. Most residents relocated farther up Bayou Lafourche to Leeville or Golden Meadow. Residents gave up on Manila Village after Hurricane Betsy in 1965, but the shrimp-drying business was already in decline and businessmen were not inclined to rebuild.

Data Source: U.S. Geological Survey (USGS) Geographic Names Information System (GNIS) 2007

Fig. 4. Post-Office Closures and Relocations

3 maps

Post offices operate in places that host a threshold population. Closure of a post office often suggests population decline. Some closures in coastal parishes are the result of consolidation with the advent of motorized rural delivery, but the general pattern of post offices moving inland mirrors the general pattern of populations closest to the water line slowly shifting inland.

Data Sources: Germann 1990; USGS GNIS 2007

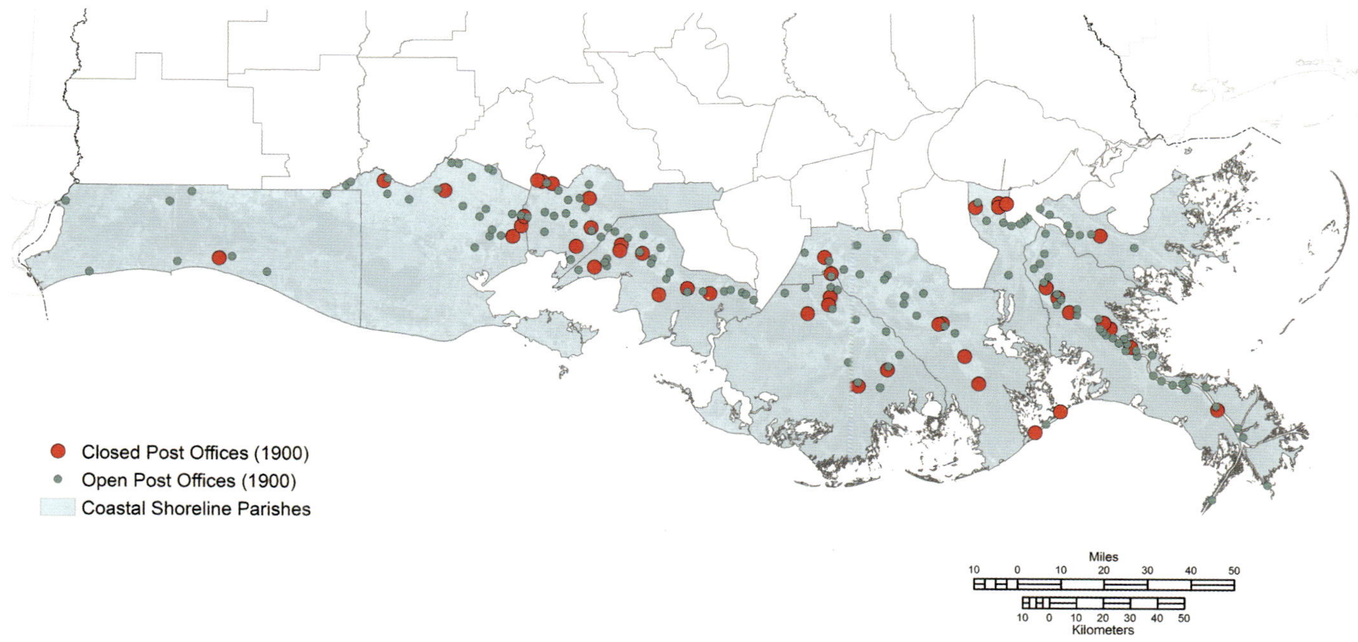

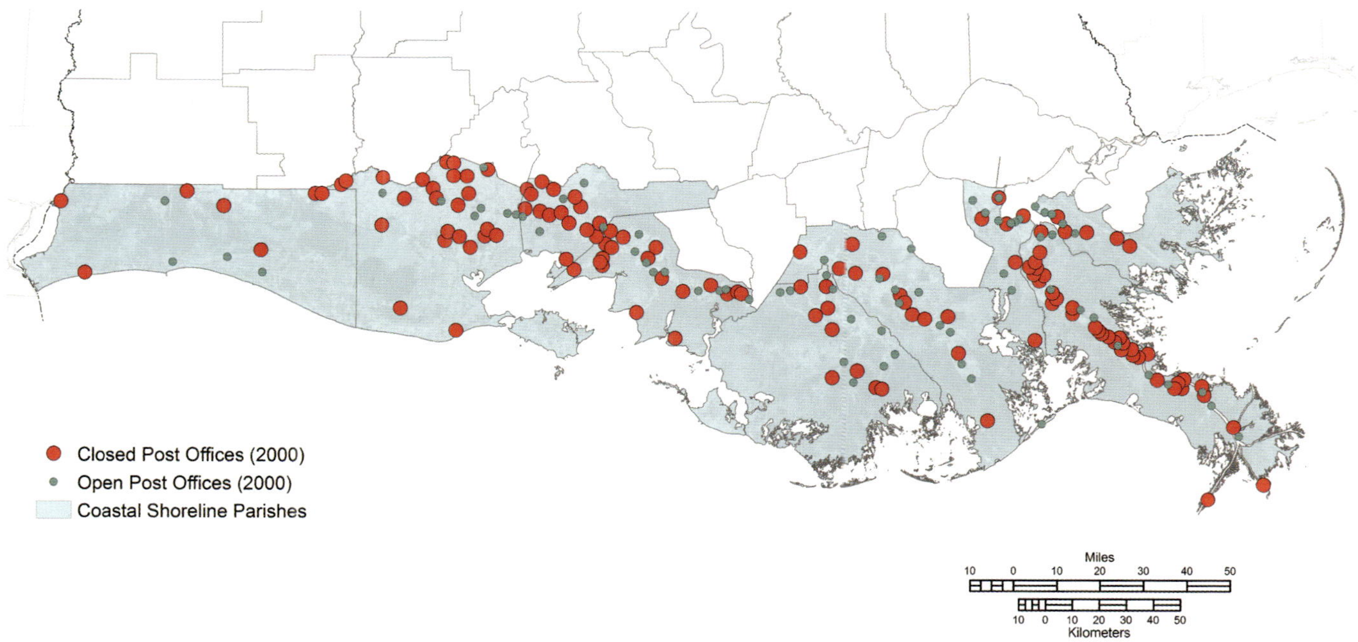

Fig. 4. Post-Office Closures and Relocations

Data Sources: Germann 1990; USGS GNIS 2007

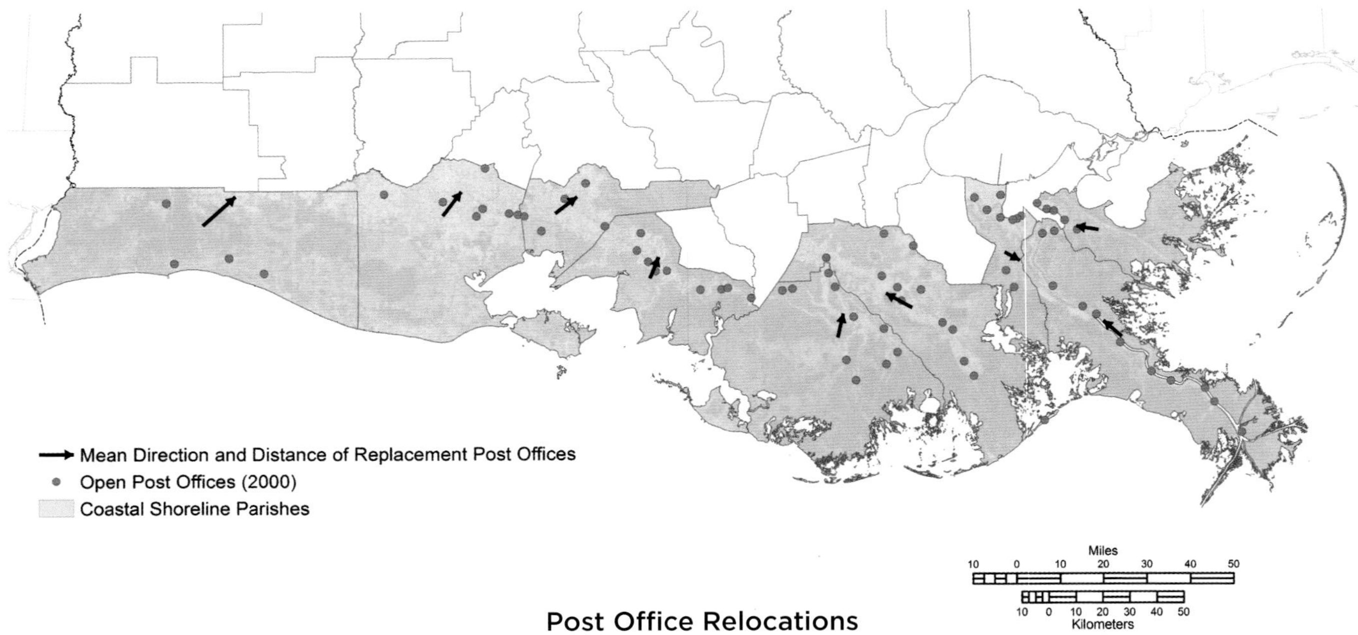

CHAPTER 1

BUILT ON UNSTABLE GROUND

Present-Day Social Vulnerability in Coastal Louisiana

LOUISIANA, LIKE OTHER COASTAL REGIONS of the United States, has seen an unprecedented level of population growth during the early years of the twenty-first century. The last two decades of the twentieth century alone saw the broader U.S. coastal population increase by 28 percent. Counties located along the Gulf of Mexico have experienced growth far exceeding this national average, with a total population increase of 45 percent during this same time period (Crosset 2004). While a combination of social and economic factors has largely driven this migration, the construction of extensive structural flood protection systems has allowed populations to move ever closer to the interface of land and water where the impacts of coastal storms and the risk of storm-induced inundation are heightened. This is nowhere more evident than in coastal Louisiana, where leveed areas reach deep into the coastal marshes and human settlements exist on narrow fingers of land at the watery fringes of civilization. Strikingly, residents of these small rural communities are often crowded together on the limited high ground in density levels approaching those of the greater New Orleans area (Laska et al. 2005).

Even as the population continues to migrate into the broader coastal zone, unprecedented natural and human-induced changes to the environment have increased the risks faced by those closest to the water. The U.S. Geological Survey has reported, for example, that coastal Louisiana has undergone a net loss of approximately 1,883 square miles of land and wetlands from 1932 to 2010 (Couvillion et al. 2011). The loss of this coastal land represents much more than merely the physical loss of ground. It is an environmental transformation with the potential to completely break down the economic and social fabric that has sustained the region for generations. For an environmental change to alter human systems in any dramatic way, it must have some impact on human well-being. This impact can be direct, such as immediate loss of life or property, or it can be indirect, affecting the social and ecological systems on which communities depend, such as agriculture or fisheries (Lenton 2013). Loss of the coastal wetlands represents the deterioration of the vital shoreline protection zone that insulates coastal communities from the impacts of storms and the deleterious effects of climate change, such as accelerated sea-level rise and increased frequency and intensity of extreme tropical weather events (Laska et al. 2005). Additionally, this zone is a vital ecological link between human communities and the benefits provided by local ecosystems. The loss of these coastal ecosystems would have long-lasting effects on the traditional economic activities, such as fishing, hunting, and trapping, that have been at the core of Louisiana's coastal culture, as well as an essential economic engine for the region and the nation at large.

According to the Millennium Ecosystem Assessment, large ecosystem-scale changes in the physical environment produce cascading consequences when viewed in terms of human-environment relations (Millennium Ecosystem Assessment 2005). This means the impacts on human communities may increase at a rate surpassing that of the triggering environmental shock. These abrupt and potentially irreversible changes to the social-ecological fabric could have dramatic impacts on the future distribution and composition of human settlements in coastal Louisiana.

Regional demographic and economic patterns in the landscape highlight the critical roles that space and place play in social vulnerability. In many cases, differential distribution patterns of social and economic vulnerability associated with environmental change may result in a greater potential for harm to particular groups within the population. Certain groups are more sensitive to change and less able to adapt effectively when faced with environmental shocks and the associated disruptions to daily life. The poor, the elderly, the young, and minority populations are the segments of society that typically face greater challenges when responding to hazards. Consequently, the social impacts of future coastal environmental change will not be evenly distributed throughout the population. How these impacts touch communities and individuals will be guided by the degree of social and economic vulnerability of the population.

Vulnerability is best thought of as the inability of a social group to respond to and overcome the effects of a hazard within a defined geographic context (Dolan and Walker 2006). If we consider social and economic vulnerability within the context of an interconnected social and ecological system, such as that found in coastal Louisiana, where coastal fisheries and other human uses of the ecosystem strongly bind human communities to the environment, then we must also consider the characteristics of exposure to any number of social and ecological stresses (Turner et al. 2003). The stronger the social and economic ties binding the community to the environment, the more vulnerable that community becomes to any alterations in the environment. Social and economic pressures for exploitation of natural resources, for example, have changed coastal ecosystem structures, and the impacts of past and present human activities will continue to accumulate (Chesney et al. 2000). Combined with continued wetland loss, communities in coastal Louisiana are predicted to experience increased potential for repeated flooding, decreased income from commercial and recreational fishing operations, a narrowing of local employment opportunities, a loss of subsistence seafood, and an overall decline in public health due to the substitution of foods of lesser quality in daily diets (Colten 2014). Burdens such as these are often borne disproportionately by residents on the lower end of the socioeconomic spectrum.

Louisiana is projected to lose an additional 1,750 square miles of land in the next fifty years if no action is taken to rebuild and protect the state's coastal wetlands (CPRA 2012). As more and more land is lost, the buffer against storm surge and other coastal hazards is reduced. The coastal hazards risk zone will continue to expand farther inland, putting an increasing number of communities and the natural resources on which they depend at risk. How these communities respond and adapt to increasing risk levels is largely a function of the various sensitivities and inherent vulnerabilities of the population and population subgroups. Further, the factors that define present-day population vulnerability in coastal Louisiana are highly variable across space. Ethnic populations are clustered rather than evenly distributed along the coast (Laska et al. 2005). Thus, they may face disproportionate risks to certain hazards. An examination of the patterns of localization and regionalization of vulnerability in coastal Louisiana will provide valuable insights for both residents and policy makers and help them to assess risk levels and determine possible ways to navigate future changes to their environment.

Louisiana's coastal zone has been variously defined depending on the specific needs of stakeholders. Legally, the coastal zone encompasses tidally influenced waters with a measurable saltwater content under normal weather conditions and adjacent shorelands. The coastal zone boundary also has a biogeophysical aspect, incorporating factors such as geology, elevation, soil and vegetation types, and salinity. Finally, there is a coastal use and human dimension to the coastal zone. The National Oceanic and Atmospheric Administration's (NOAA) Coastal Assessment Framework has defined Coastal Watershed Counties as those counties that have a substantial watershed-based impact on coastal and ocean resources. Coastal Shoreline Counties, on the other hand, are those counties directly adjacent to the open ocean and major estuaries. These counties host the majority of the economic production associated with coastal resources but also bear much of the direct effects of coastal hazards. This atlas will focus largely on the coastal shoreline and coastal watershed parishes of the state, while utilizing other scales of analysis when broader or narrower data patterns require analysis.

POPULATION DISTRIBUTION

Louisiana's coastal zone offers an abundance of natural resources and environmental amenities that have drawn residents down the bayous and waterways of the region for centuries. Yet the very waters that have sustained these populations, their culture, and their communities have served as a limiting factor in coastal development. Coastal Louisiana's populations reside on narrow threads of land, threads that grow thinner and thinner as the land transitions to water. The geography of the region that restricts where communities can locate also serves to enhance the vulnerability of coastal communities. Proximity to the open water increases the risks of tropical weather impacts and coastal flooding. The limited land area also restricts the construction of road networks, which in turn limits options for evacuation during storm events.

Fig. 5. Population Density in Coastal Louisiana (2010)

Using a combination of census block data and land-cover datasets, dasymetric mapping techniques highlight population density patterns that are distorted in the census geographical units.

Data Sources: USGS National Land Cover Database; U.S. Census Bureau

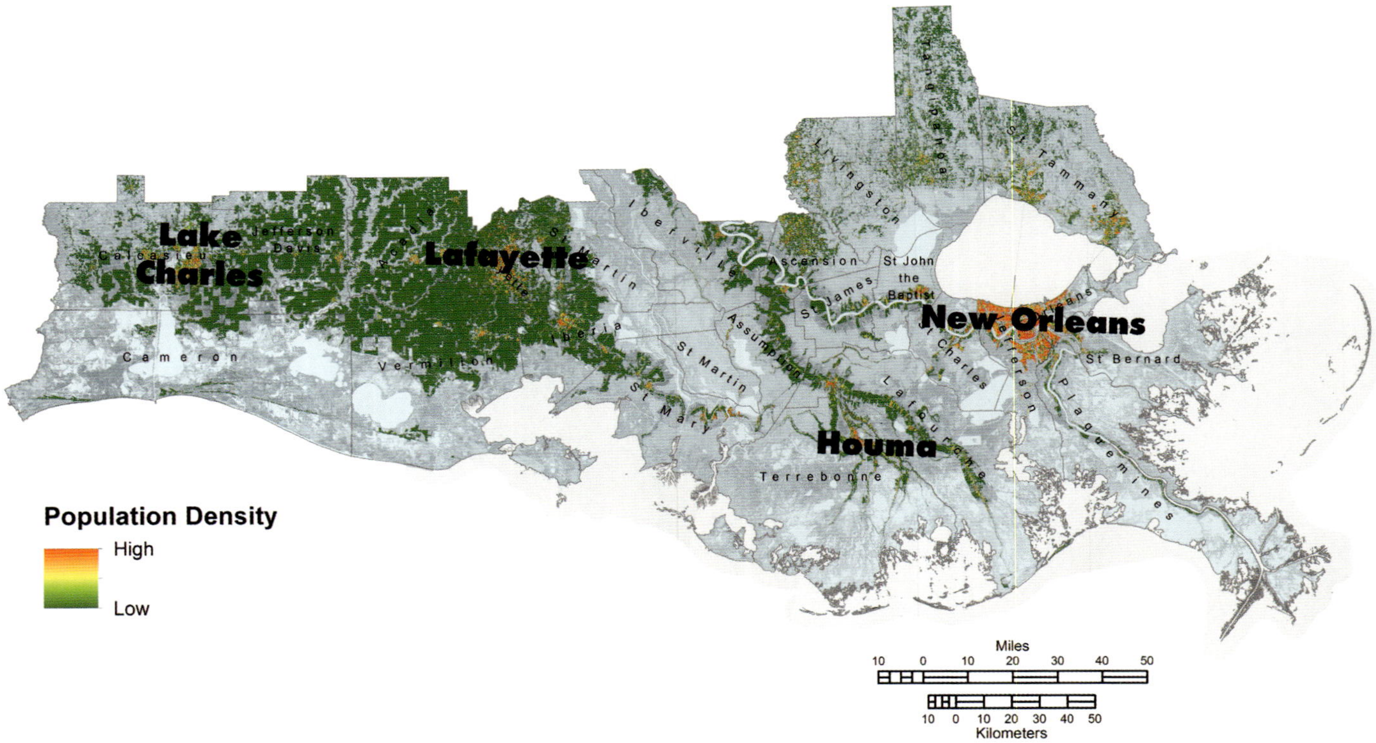

Population density patterns in southwest Louisiana differ considerably from those of the deltaic southeast (fig. 5). In most cases, however, these communities tend to cluster atop the limited high ground and in linear patterns around major roadways. Communities located far afield from major road networks tend to be smaller and more dispersed and isolated, especially as one moves farther inland. Overall, the density of development in the deltaic southeast is much greater than in the southwestern chenier plan.

Fig. 6. **Population Density in Lake Charles, Louisiana (2010)**

Prepared using dasymetric mapping techniques.

Data Sources: USGS National Land Cover Database; U.S. Census Bureau

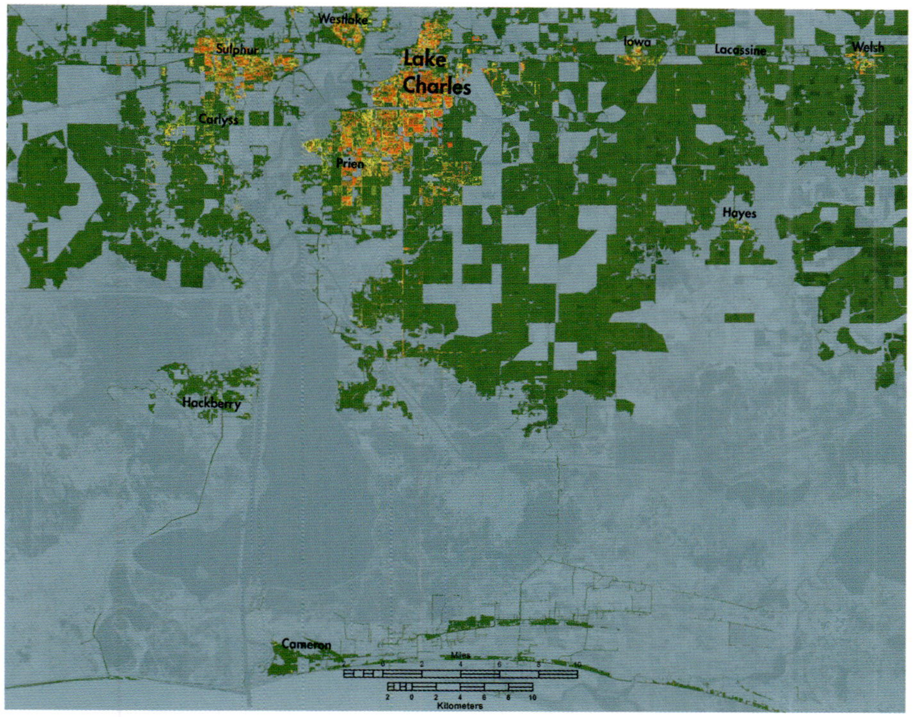

Fig. 7. **Population Density in Lafayette, Louisiana (2010)**

Prepared using dasymetric mapping techniques.

Data Sources: USGS National Land Cover Database; U.S. Census Bureau

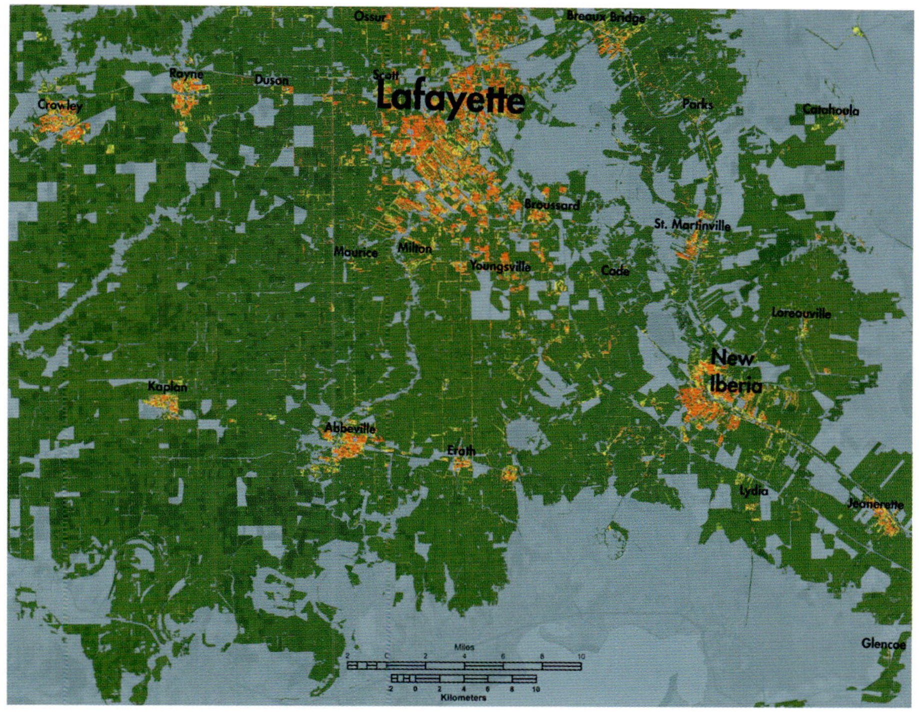

While still characterized by extensive coastal marshland, the limited high ground in coastal southwest Louisiana is located on relict beach ridges, or cheniers, running parallel to the coastline, effectively disconnected from the populated high ground farther inland (figs. 6 and 7). Without that direct connecting thread of high ground stretching inland, population development has historically been limited in coastal southwest Louisiana.

Fig. 8. Population Density in Houma, Louisiana (2010)

Prepared using dasymetric mapping techniques.

Data Sources: USGS National Land Cover Database; U.S. Census Bureau

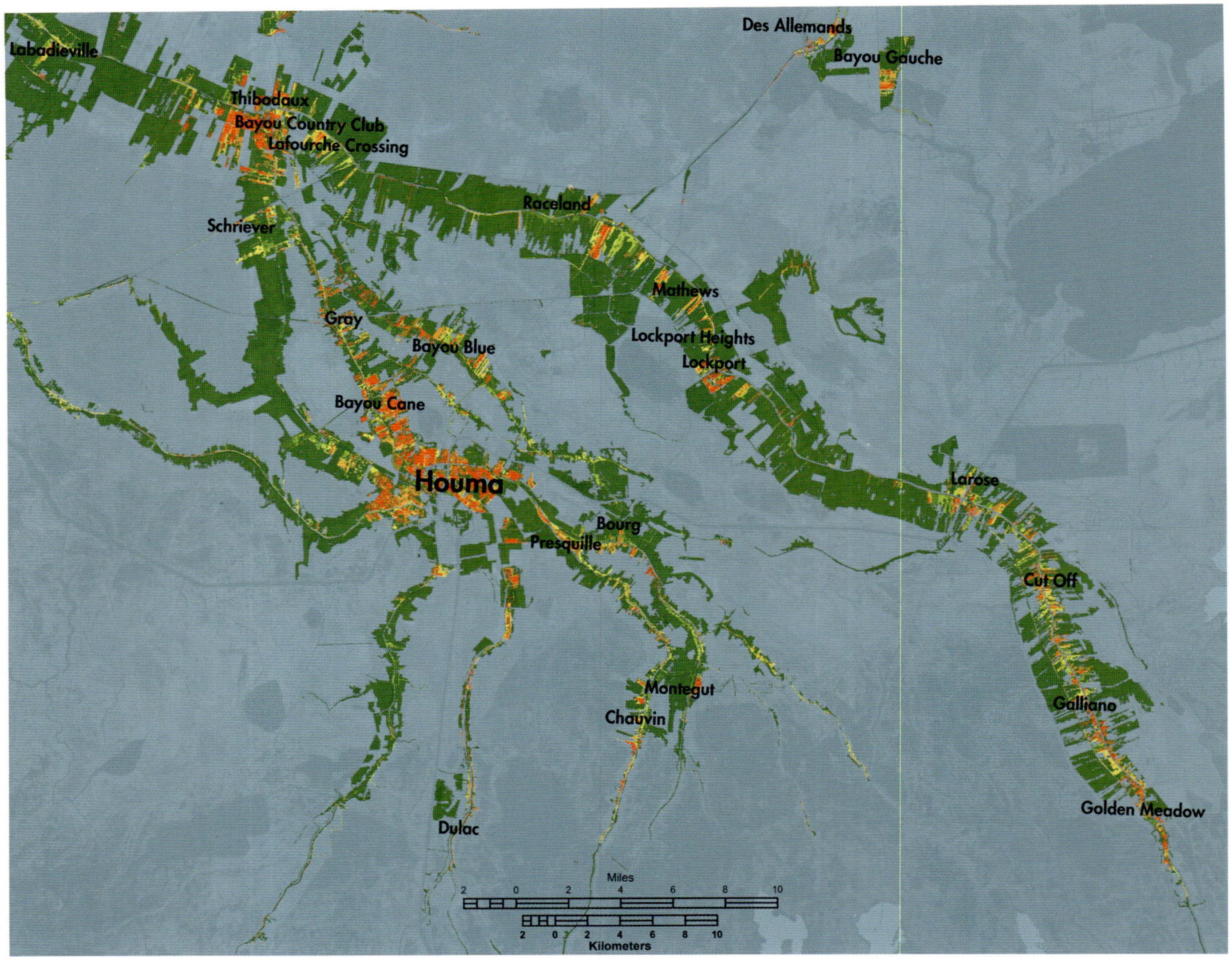

The density of development on the limited high ground and natural levees in the region has in some cases approached the density of inland urban areas. This is particularly true of development in the deltaic plain of the Mississippi River, extending from the edge of the chenier plains eastward (figs. 8 and 9), where human development primarily encompasses those natural ridges located along the remnants of many former Mississippi River delta complexes. This settlement pattern presents numerous challenges for coastal residents.

Fig. 9. **Population Density in New Orleans, Louisiana (2010)**

Prepared using dasymetric mapping techniques.

Data Sources: USGS National Land Cover Database; U.S. Census Bureau

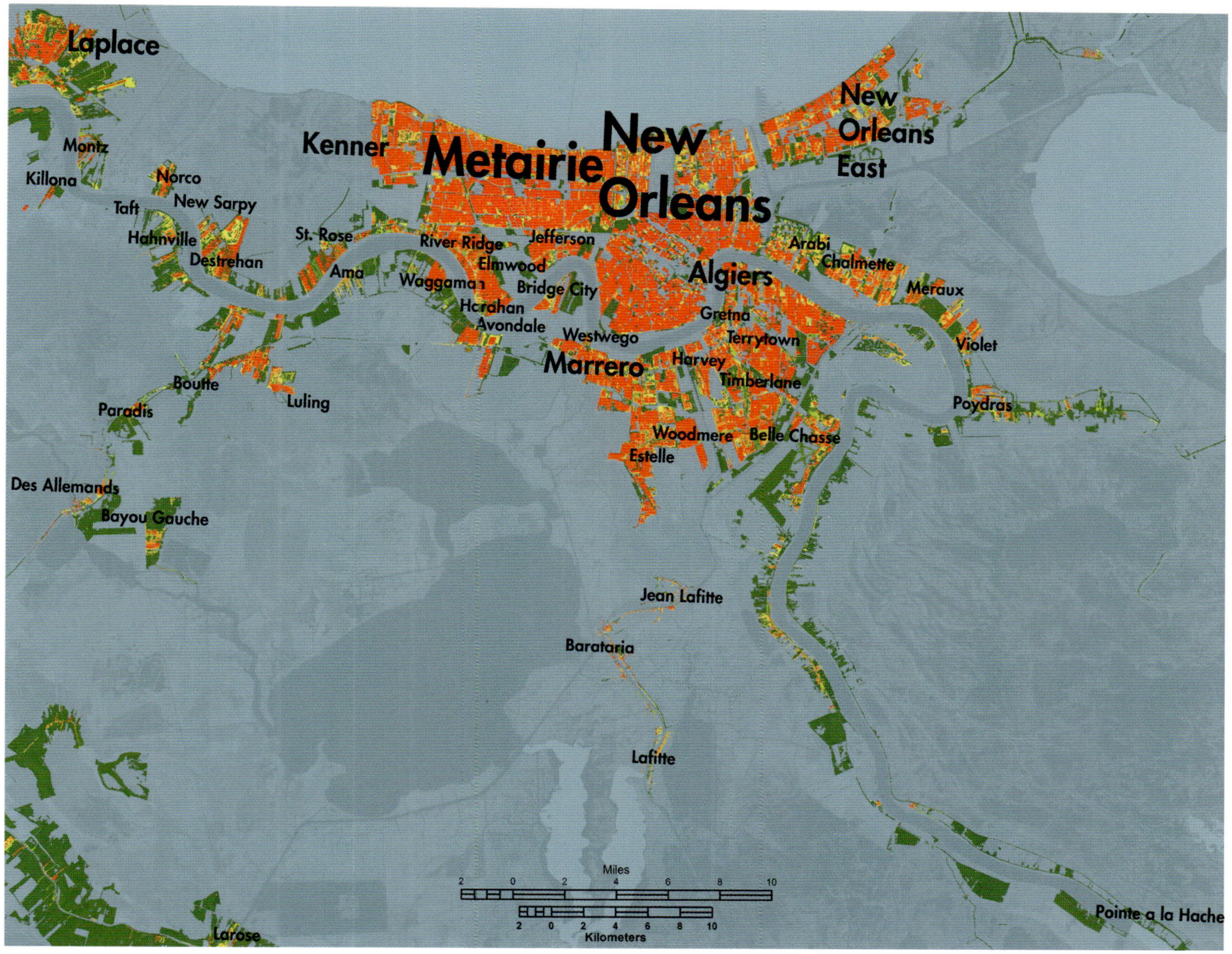

Even large urban areas such as the New Orleans metropolitan region tend to initially develop on the limited high ground. As urban growth progresses and coastal engineering structures such as flood protection levees are built to protect residents, settlement patterns begin to take the form of these coastal protection features, with populations becoming ever more densely concentrated within the protection levees (fig. 9).

Fig. 10. Total Population of Communities in Coastal Shoreline and Coastal Watershed Parishes (2010)

Data Source: U.S. Census Bureau

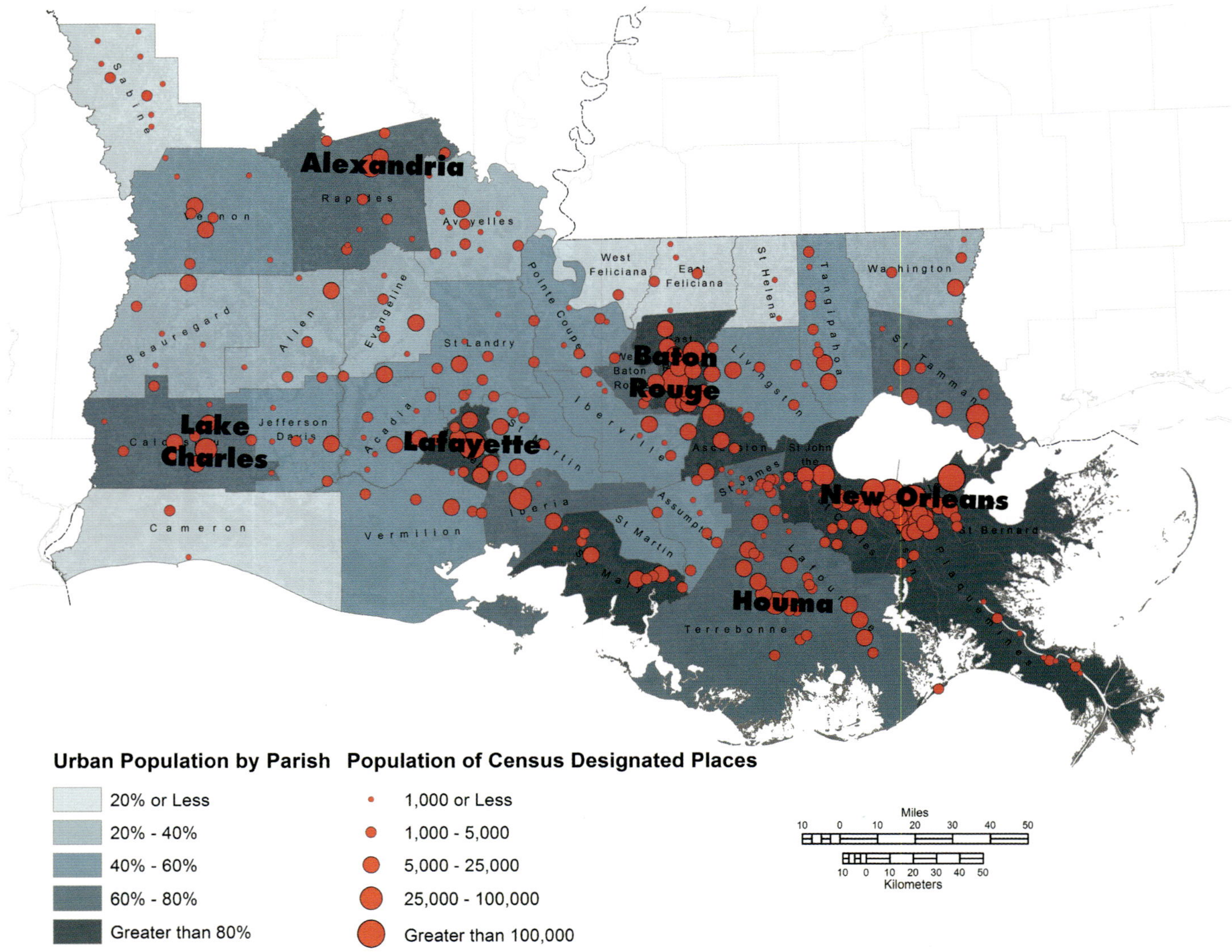

Urbanization at the parish level tends to develop linearly as well, with urbanized core parishes connected to one another by less urbanized, but still highly developed, urban corridor parishes (fig. 10). In coastal Louisiana, this pattern is apparent where interstate highways and a handful of larger state highways connect all of the highly urbanized parishes in the southern tier of the state.

NATURAL RESOURCE DEPENDENCE

The degree to which communities rely on natural resources for employment and, in some cases, sustenance is directly correlated to their vulnerability to environmental changes. Louisiana's working coast depends heavily on natural resource extraction—both renewable, in the form of fisheries, agriculture, and forestry, and nonrenewable, in the form of oil and gas production. There are clear patterns in the distribution of natural resource employment in coastal Louisiana specific to each of these industries. Obviously, physical characteristics of the environment are the primary determinants of natural resource clusters. However, advances in technology, whether they be new irrigation techniques, fishing gear types, or oil extraction equipment, also play a role in determining where natural resource centers are located.

Fig. 11. Regional Clusters of Natural Resource Employment

Urban communities have significantly less natural resource employment (less than 2.58 standard deviations), while rural areas have significantly more (greater than 2.58 standard deviations).

Data Source: U.S. Census Bureau

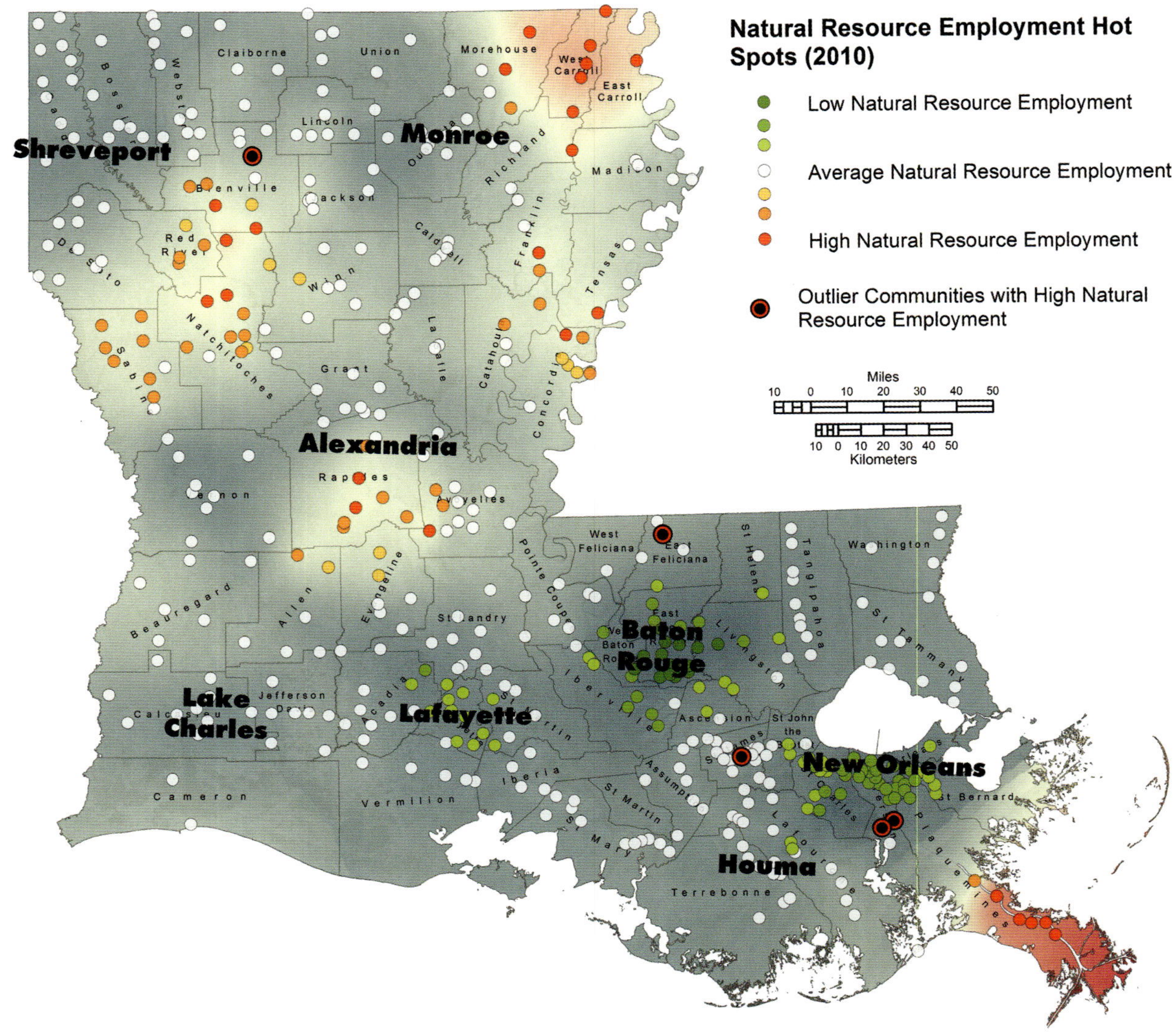

In Louisiana, several agricultural employment centers can be found in the central and northern portions of the state (fig. 11). Employment clusters related to fisheries occur in the coastal zone and the Atchafalaya Basin. The most significant regional grouping of fishing-related communities can be found on the Barataria Basin in southeast Louisiana. This area is highly invested in the fishing/seafood industry and includes many of the communities located on the lower Mississippi River in Plaquemines Parish, as well as the communities of Lafitte, Jean Lafitte, and Barataria, situated on the northern edge of the Barataria Basin in Jefferson Parish.

Fig. 12. Farming and Agriculture in Coastal Shoreline and Coastal Watershed Parishes (2010)

4 maps

Data Sources: U.S. Census Bureau; USDA, National Agricultural Statistics Service

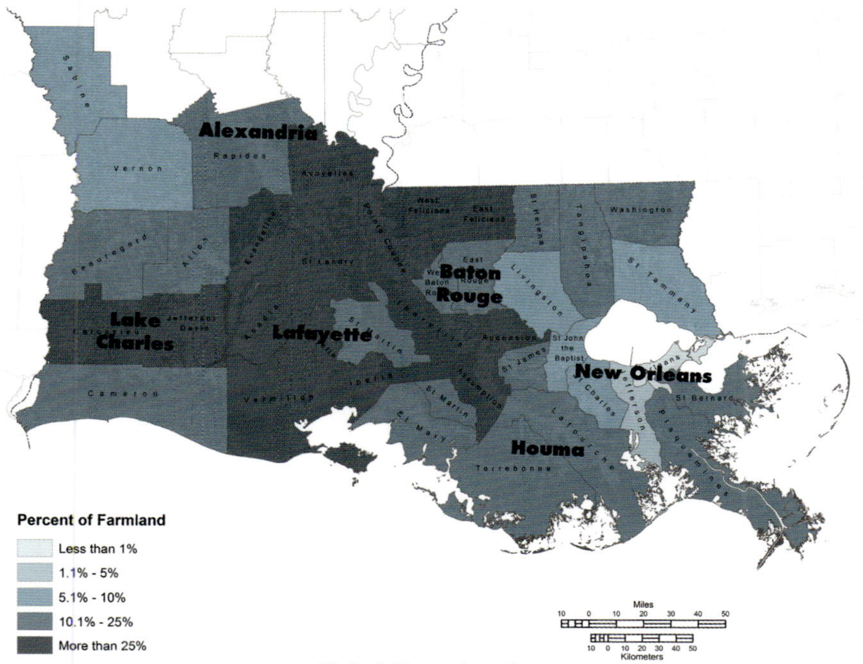

Total Farmland

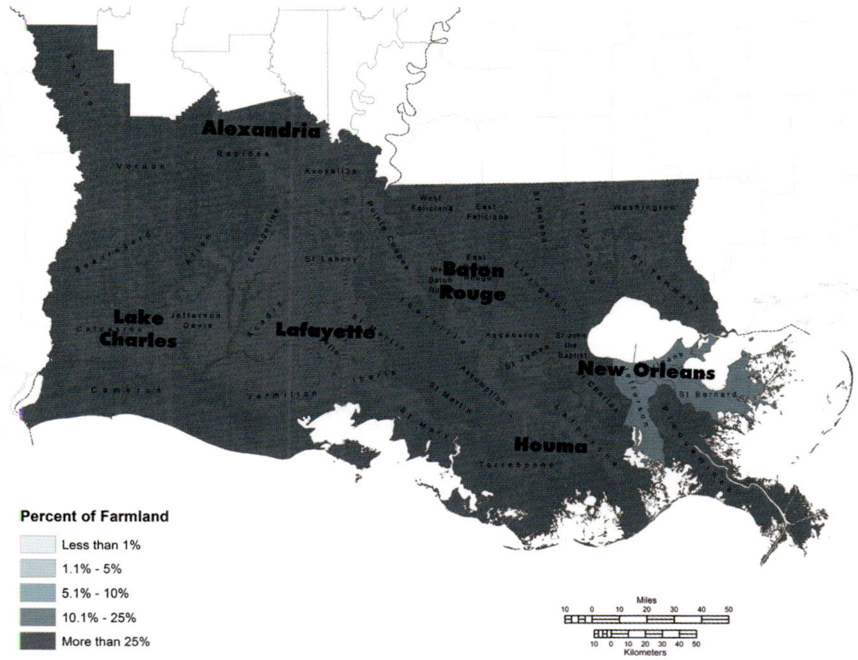

Cattle

On a broad parish level, the regionalization of agriculture becomes apparent (fig. 12). The distribution of cattle farms is the most dispersed, with a preponderance of ranches stretching along the Sabine River on the western edge of the state, in the Florida Parishes north of Lake Pontchartrain, and in the area around Lafourche Parish in south Louisiana. Crops such as rice and sugar show a much higher degree of regionalization, with rice production focused in the southwestern parishes and sugar in the south-central parishes, particularly in the rural areas between Baton Rouge, Houma, and Lafayette.

Fig. 12. Farming and Agriculture in Coastal Shoreline and Coastal Watershed Parishes (2010)

Data Sources: U.S. Census Bureau; USDA, National Agricultural Statistics Service

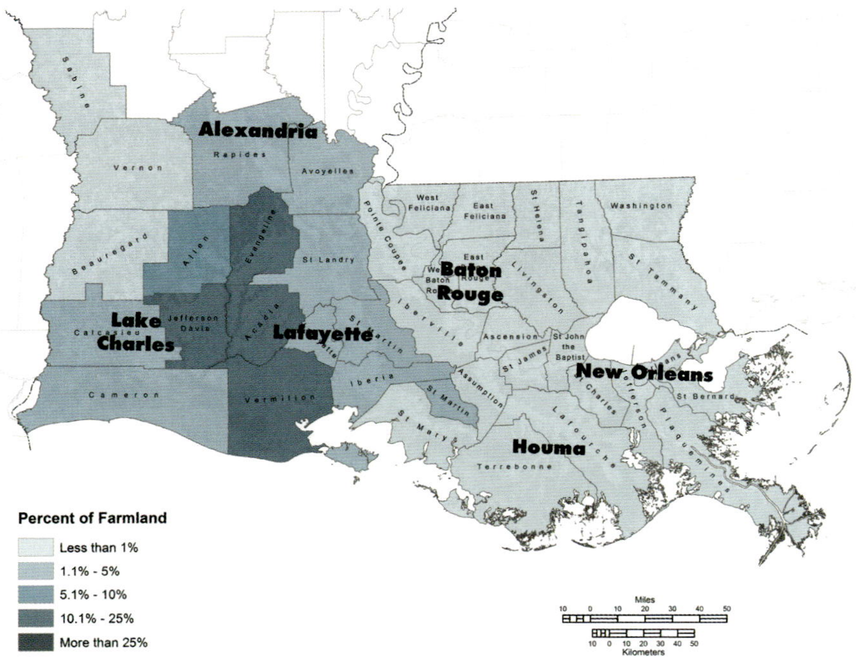

Rice

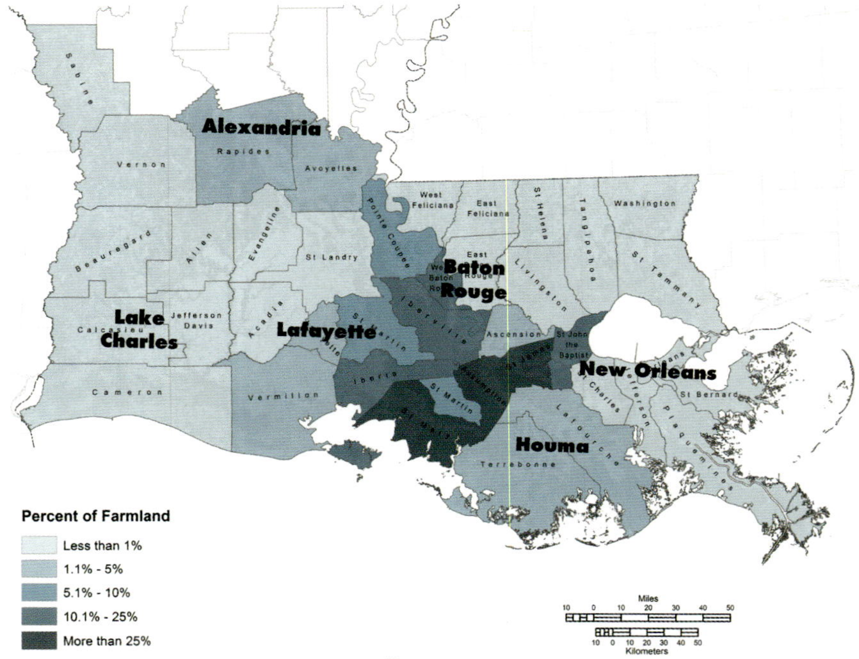

Sugar

Fig. 13. Total Value of Fisheries and Aquaculture in Coastal Shoreline and Coastal Watershed Parishes (2010)

3 maps

Data Source: LSU AgCenter

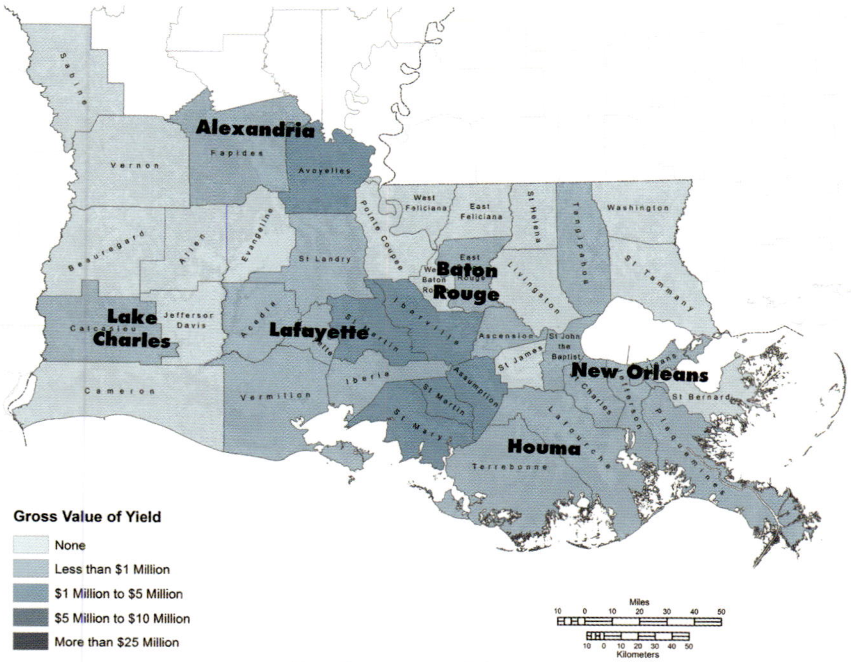

Freshwater Fisheries

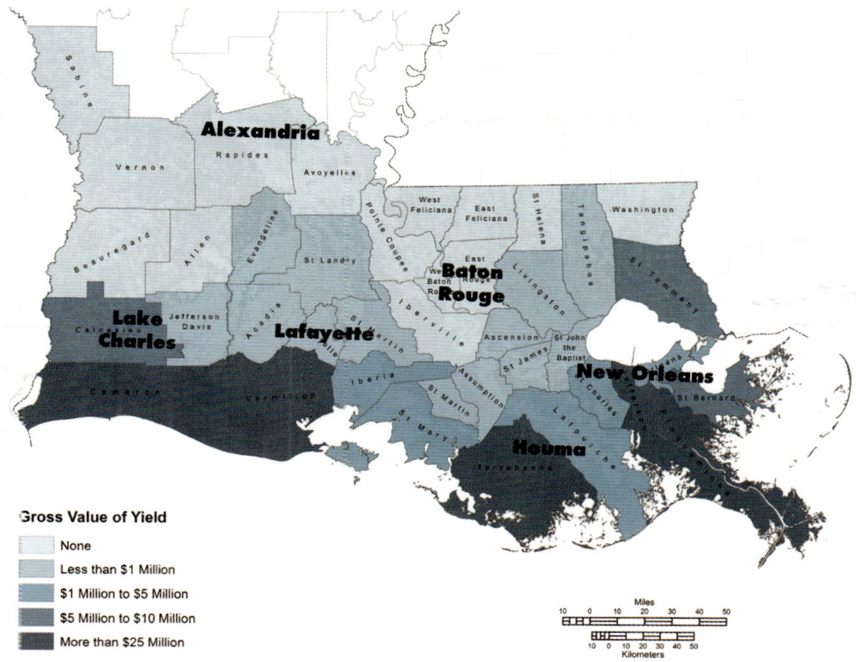

Marine Fisheries

Fisheries also show a great deal of geographic specialization, as would be expected (fig. 13). The Atchafalaya Basin, the nation's largest river swamp ecosystem, is the center of freshwater fisheries activities, whereas the marine fisheries industry dominates the coastal shoreline parishes.

Fig. 13. Total Value of Fisheries and Aquaculture in Coastal Shoreline and Coastal Watershed Parishes (2010)
Data Source: LSU AgCenter

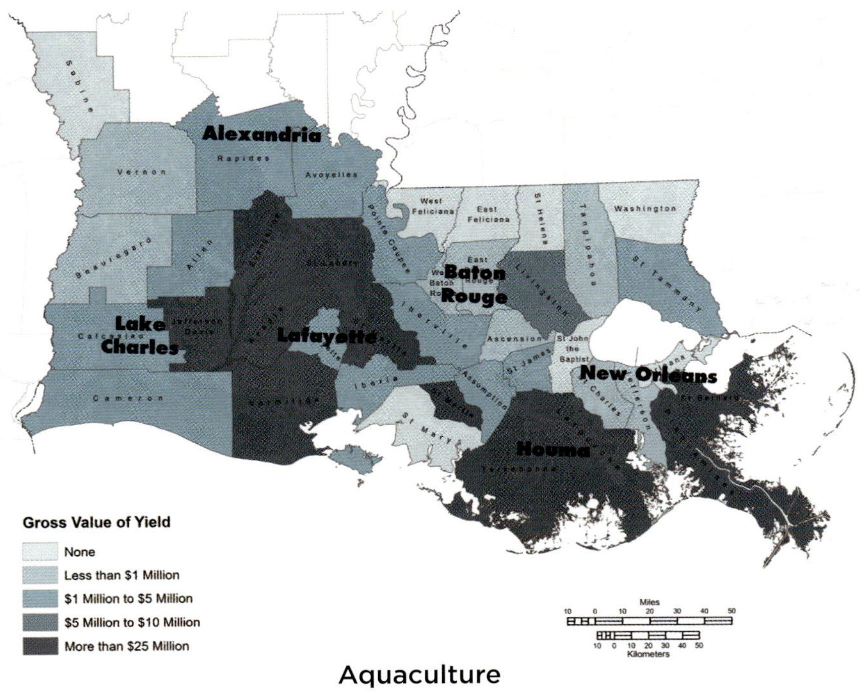

Lastly, aquaculture, or the "farming" of fish and other aquatic species, is an important industry across the coast, particularly in southeast Louisiana and in the region of southwest Louisiana around Acadia Parish, an area dominated by crawfish farming. Products of commercial aquaculture in Louisiana include finfishes, such as channel catfish, baitfish, tilapia, and recreational sport fishes; crustaceans, including freshwater crawfish and soft-shell blue crabs; and reptiles, including the American alligator and red-eared turtle.

Fig. 14. Average Annual Board Feet of Timber Produced (2007–2010)

2 maps

Data Sources: Louisiana Forestry Commission; Louisiana Department of Agriculture and Forestry

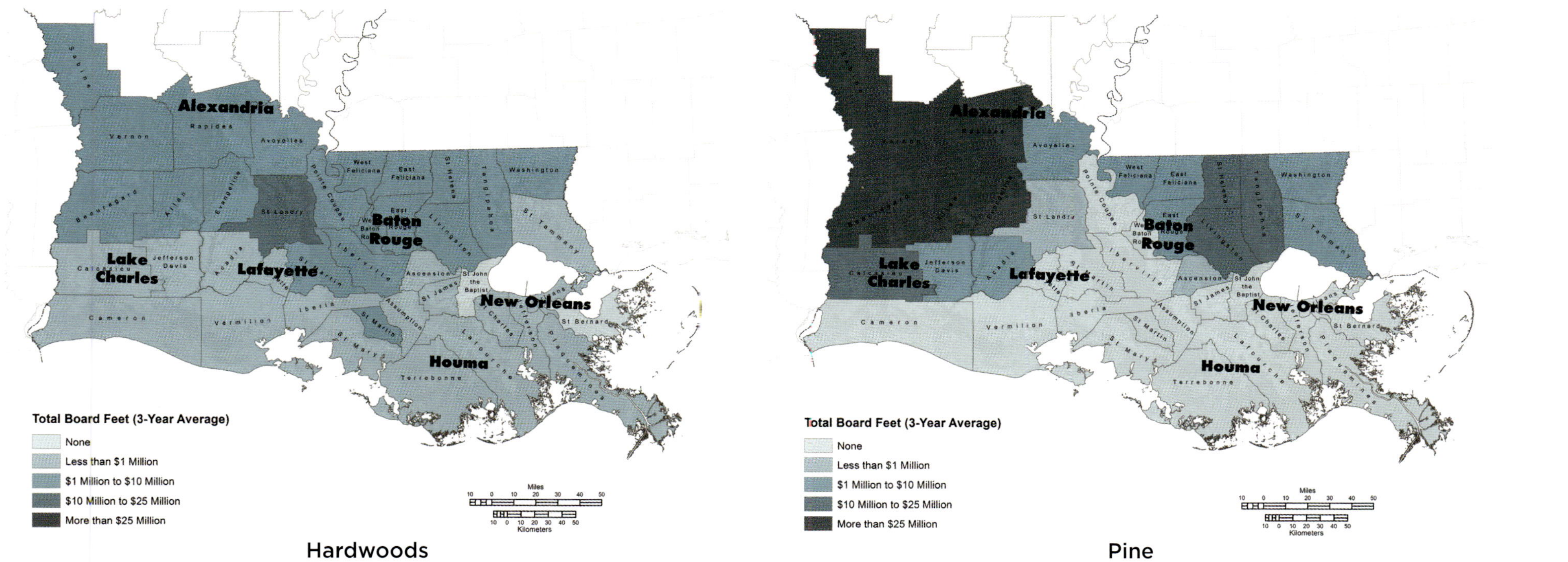

Almost all of the parishes in Louisiana's coastal zone have some degree of timber production (fig. 14). Cypress production, once the dominant forest-related industry in coastal Louisiana, collapsed in the first half of the twentieth century. Despite this, the timber industry has persisted, utilizing some forty different native timber species. While not all native species are commercially viable, when examined in total, the value of Louisiana hardwoods and pine cannot be overlooked. Pine production occurs primarily on lands outside of the coastal zone, specifically, in the Florida Parishes north of Lake Pontchartrain and in the western portion of the state. While these areas also support hardwood production, most hardwoods are drawn from the northern reach of the Atchafalaya River, such as St. Landry Parish.

Fig. 15. Density of Oil and Natural Gas Wells across Louisiana and on the Outer Continental Shelf (2010)

Data Sources: Louisiana Department of Natural Resources; U.S. Bureau of Ocean Energy Management

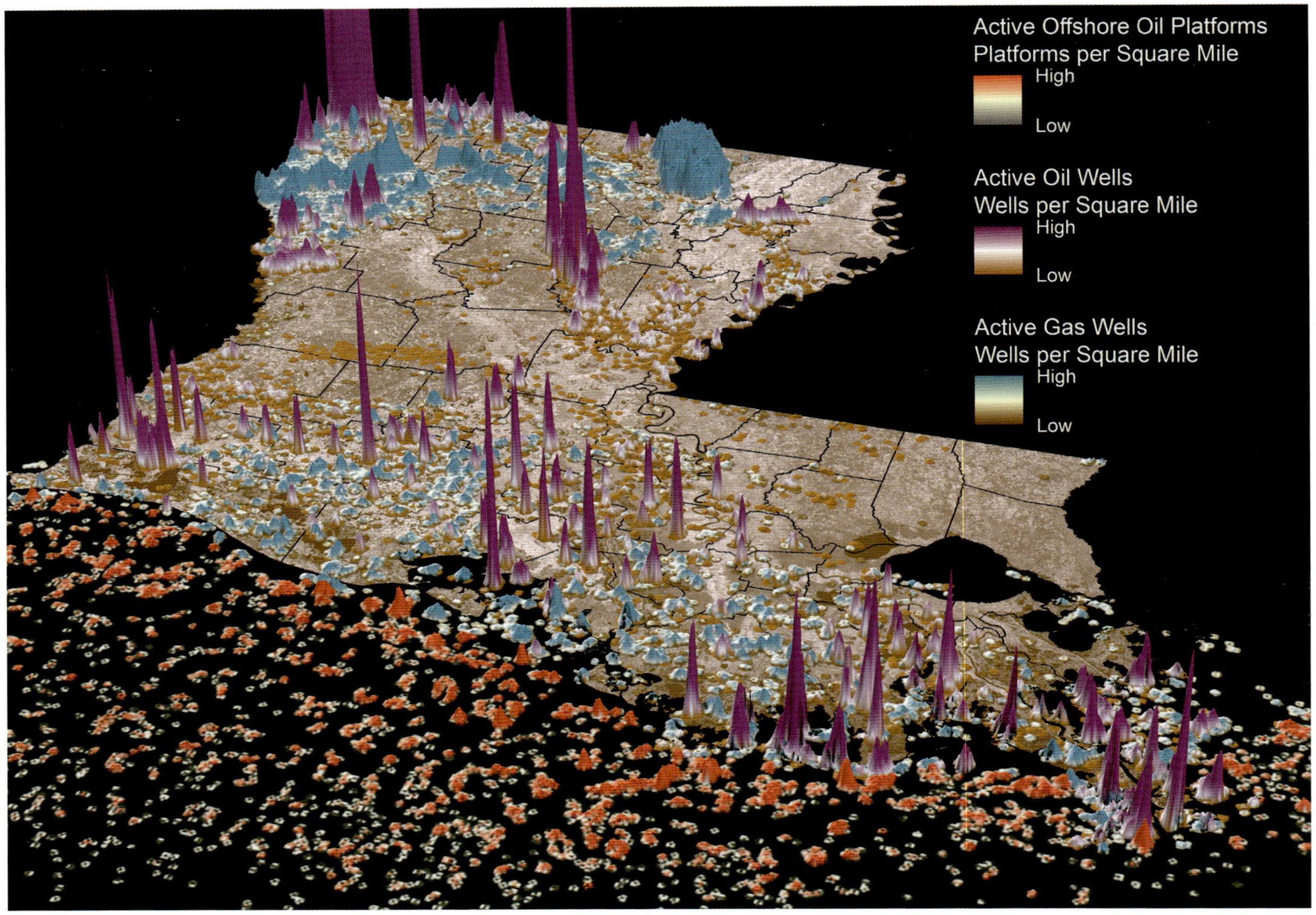

The other dominant resource-related industry in Louisiana is oil and natural gas extraction and processing (fig. 15). Oil and gas production occurs in both north and south Louisiana. Oil wells are concentrated in productive fields, while gas wells show a more widespread pattern. Outer Continental Shelf oil extraction activities in the Gulf of Mexico, as evidenced by the sheer number of offshore oil platforms, center off the coast of Louisiana and represent the extension of mineral extraction offshore.

Fig. 16. Portion of the Population Employed by Oil and Natural Gas Extractive Industries in Louisiana Communities (2010)
Data Source: U.S. Census Bureau

Fig. 17. Portion of the Population Employed by Oil and Natural Gas Extractive Industries in Louisiana Coastal Parishes (2010)
Data Source: U.S. Census Bureau

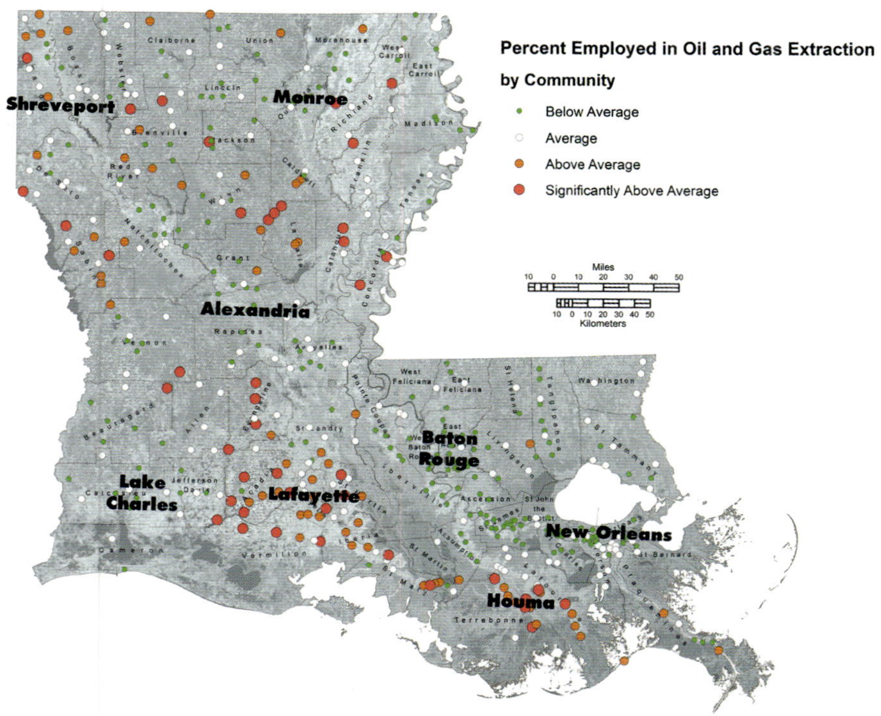

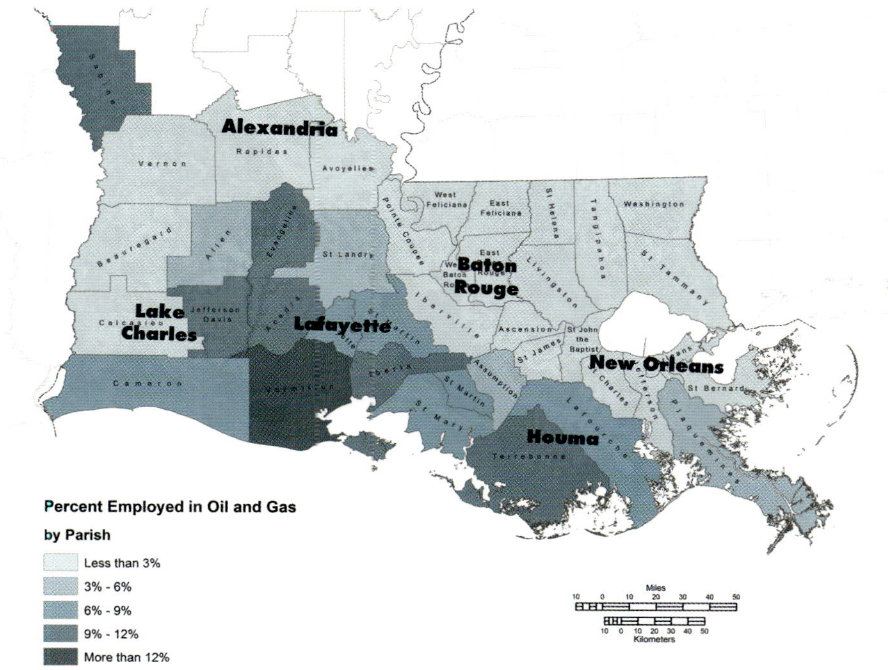

Overall, the population employed in oil and gas extraction closely mirrors the location of oil and gas fields (fig. 16). While much of this employment is directly related to the process of extracting the oil and gas, other petroleum exploration, transport, and processing activities are also prominent employers in communities across Louisiana. Notably, however, many communities in coastal Louisiana also serve as embarkation points for those employed on rigs in the deep waters of the OCS (fig. 17).

Fig. 18. Water Transportation Businesses (2010)
Data Source: U.S. Census Bureau

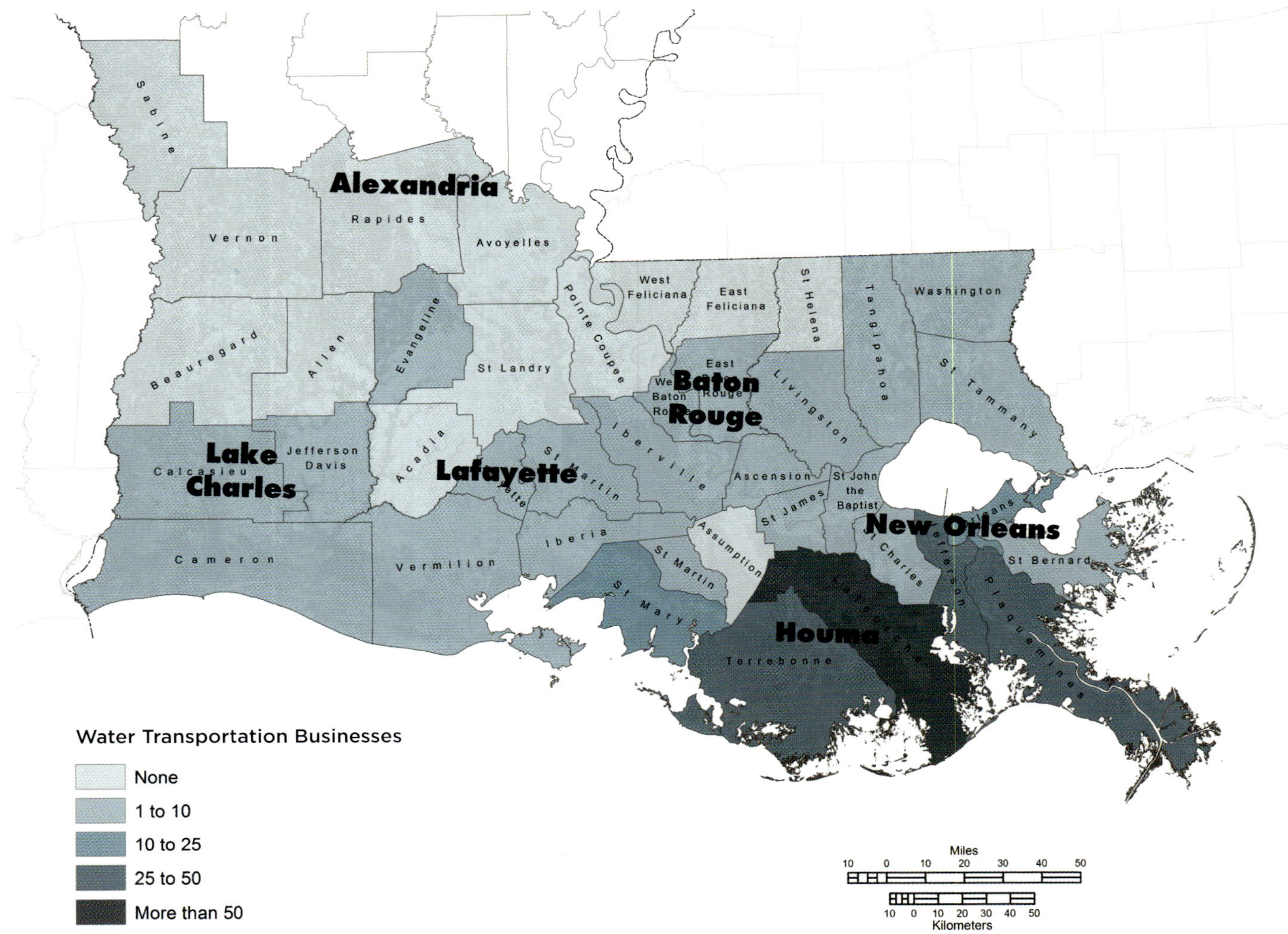

Finally, it is important to note that there is another industry that relies heavily on the environment of coastal Louisiana, though not for resource extraction. That is the water transportation and navigation industry. Virtually every parish in the coastal zone is home to a number of water transportation businesses (fig. 18). The vast majority of these businesses operate in southeast Louisiana, particularly in Lafourche and Terrebonne parishes, principal locations for the offshore energy industry, and Plaquemines Parish, with its location at the mouth of the Mississippi River. These businesses rely heavily on the navigable waterway network that crisscrosses the coastal zone, providing a valuable connection between inland areas and the coast.

Fig. 19. Public Boat Launches and Navigable Waterways

Data Sources: Louisiana State Land Office; U.S. Department of Transportation

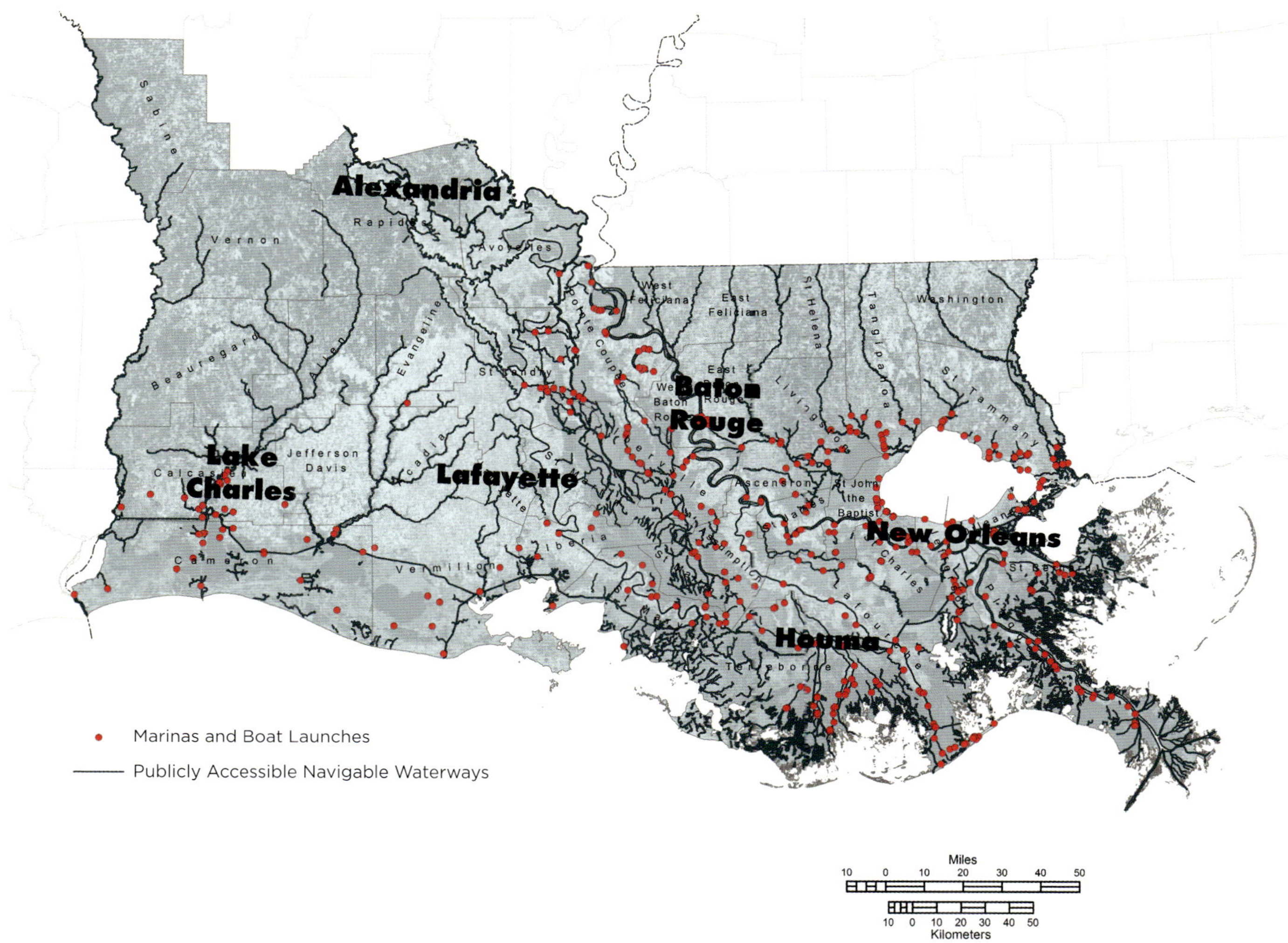

Waterways that were navigable when Louisiana attained statehood in 1812 are owned by the citizens of the State of Louisiana. Many of these waterways are publicly accessible for commercial and recreational uses. A number of private and public marinas (fig. 19) have developed in the coastal zone and beyond, allowing residents to more easily access these public waterways.

Fig. 20. Ports and Commercially Navigable Waterways

Data Sources: U.S. Army Corps of Engineers; U.S. Department of Transportation

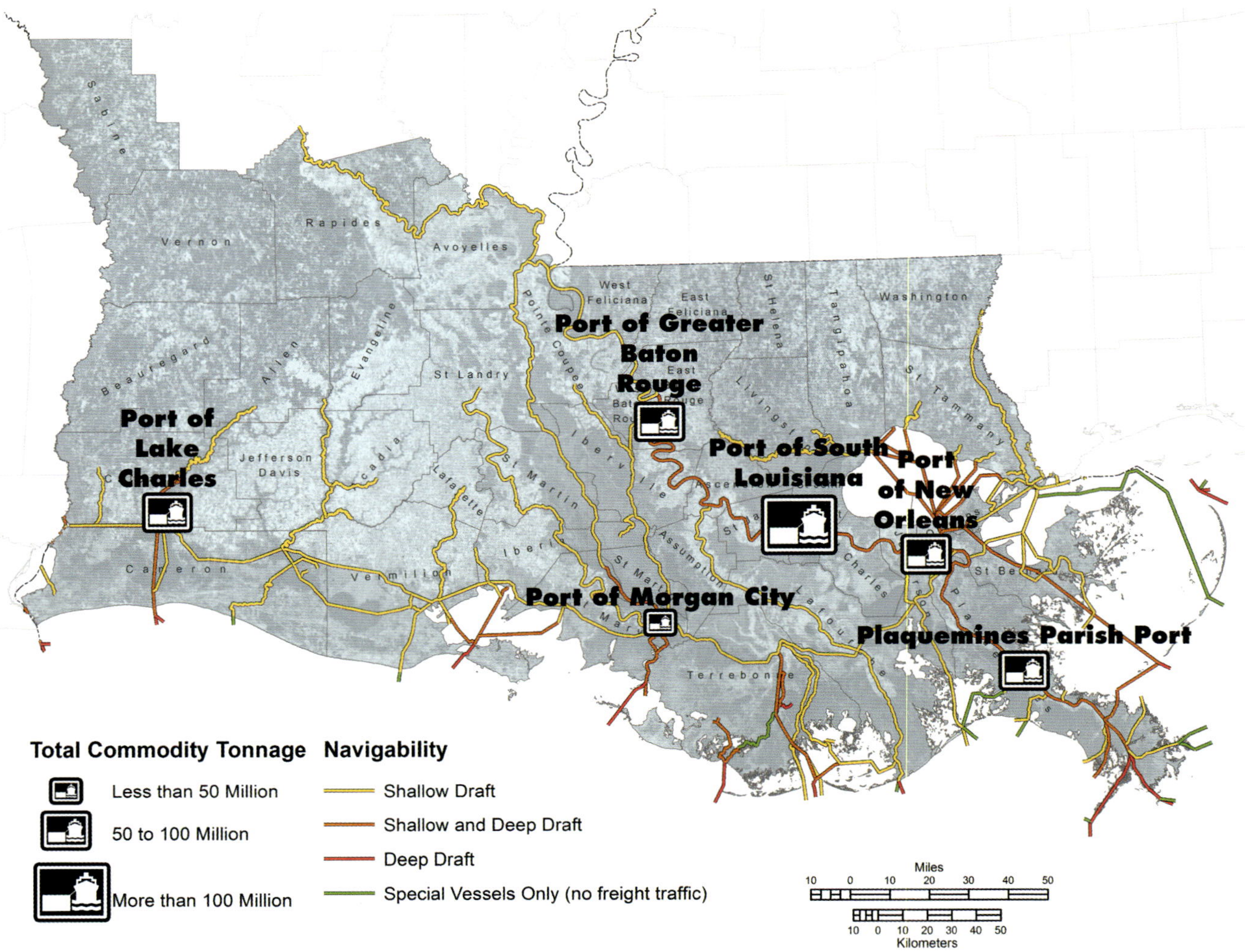

Much of the waterborne commerce to and from the midwestern United States flows through Louisiana's ports (fig. 20). These ports transfer products from barges to oceangoing vessels for export and from ships to barges for import. In total, some 19 percent of the nation's waterborne commerce passes through Louisiana's ports.

SOCIAL AND ECONOMIC VULNERABILITY

Social vulnerability is based on the idea that particular population subgroups have innate characteristics that impact their ability to respond effectively to environmental shocks and stresses. Underserved populations often do not have access to the resources that would allow them to effectively anticipate, cope with, and recover from a natural or technological hazard. Lack of income, lack of transportation, age, gender, and minority status all influence the ability of residents to respond to crises. Many of these factors are highly co-dependent, interacting with and reinforcing one another. Economic status, for example, interacts with most other social factors to reinforce the vulnerability (or lack thereof) of a particular social group.

Fig. 21. Income and Poverty in Coastal Shoreline and Coastal Watershed Parishes (2010)

3 maps

Data Source: U.S. Census Bureau

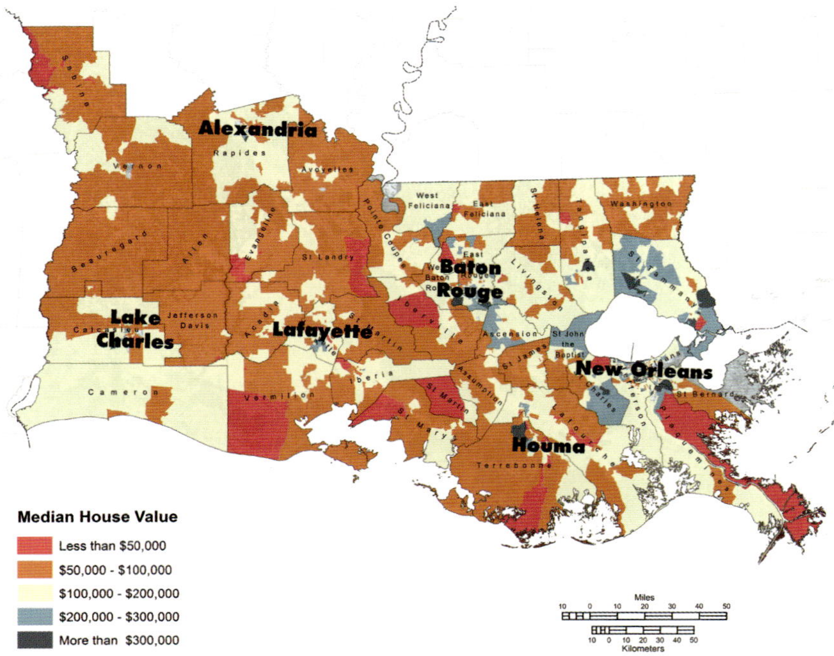

House Value

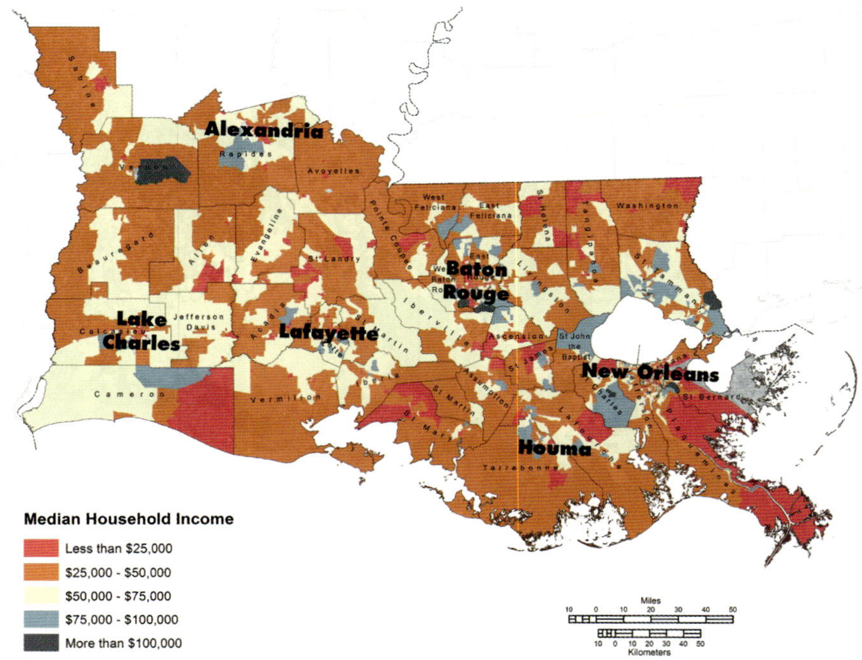

Income

Wealth and income levels directly affect the ability of individuals and communities to effectively respond to risk. The percentage of the population living in poverty is a direct measure of the community's difficulty in evacuating and finding shelter when faced with a natural hazard event. There is a dichotomous situation in coastal Louisiana, where many residents are low income but do not fall beneath the poverty line (fig. 21). This creates conditions where individuals or families are more vulnerable to any economic shock that could drive them into poverty.

Fig. 21. Income and Poverty in Coastal Shoreline and Coastal Watershed Parishes (2010)

Data Source: U.S. Census Bureau

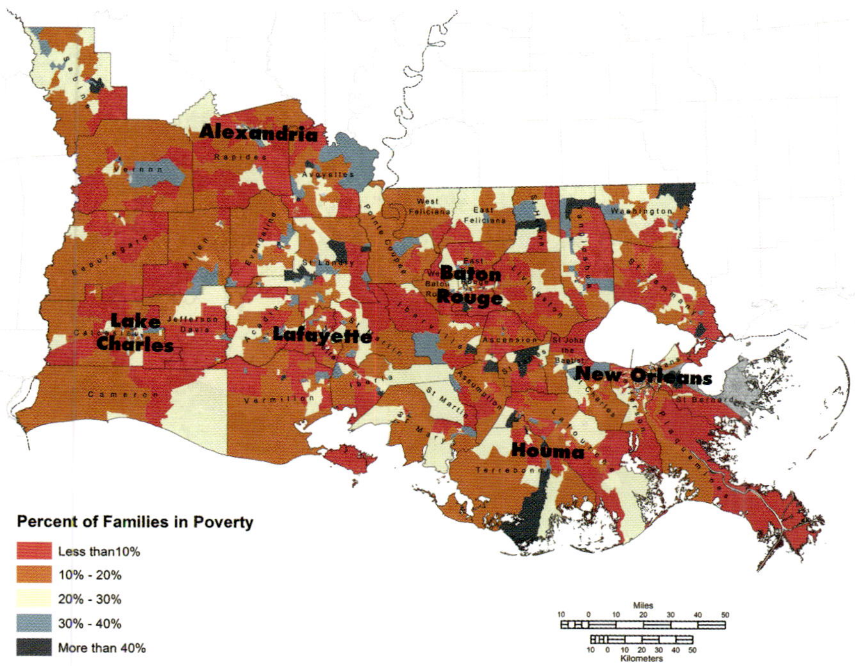

Poverty

Fig. 22. Distribution of Minority Populations by Census Block (2010)

4 maps

Data Source: U.S. Census Bureau

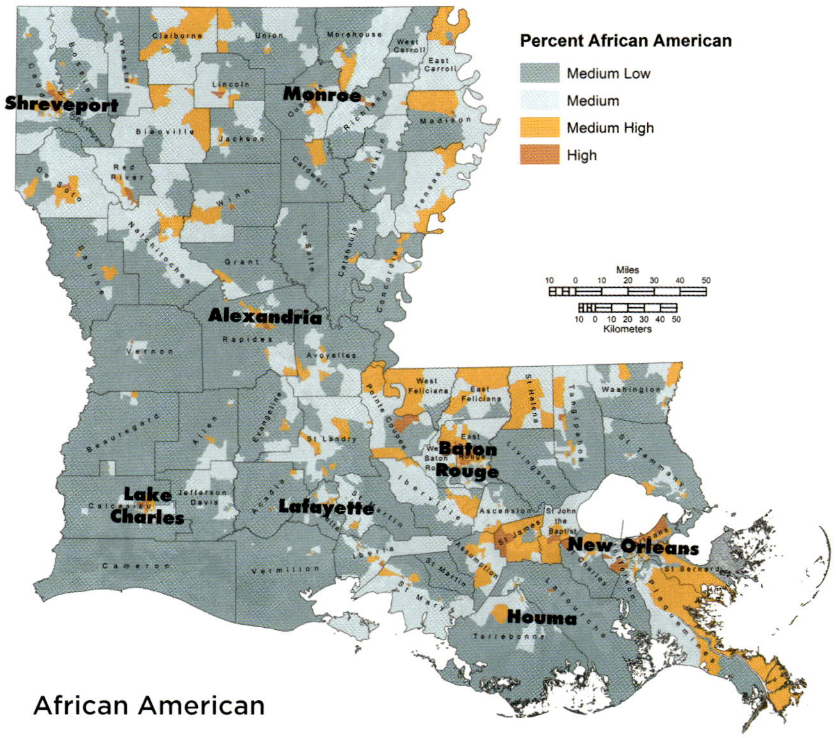

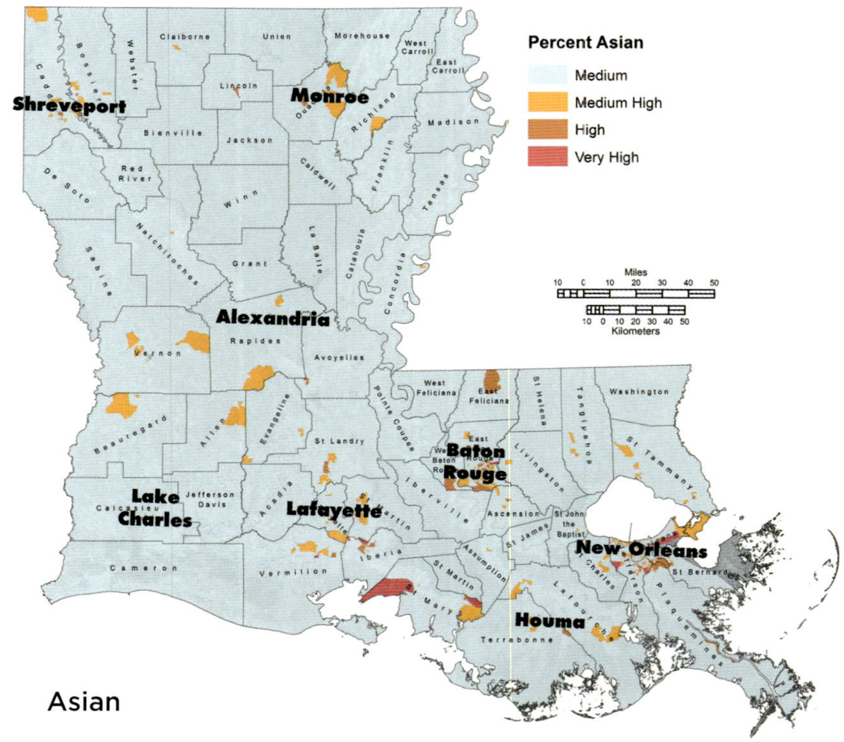

Race and ethnicity are also viewed as key vulnerabilities for a number of reasons. Minority residents are often socially marginalized and have limited access to social and political resources. This marginalization is compounded when examined in tandem with high rates of poverty. In coastal Louisiana, minority residents tend to cluster in urban areas (fig. 22). This is especially true of the African American, Asian, and Hispanic populations. There are many exceptions to this generalization. Many African Americans reside in rural agricultural communities in the Florida Parishes and along the Mississippi River. Similarly, populations of Vietnamese fishermen occur in Plaquemines Parish, and Hispanic Isleño populations in rural St. Bernard Parish. Unlike the other population groups mentioned, Native Americans in south Louisiana reside largely in the coastal zone, in southern Lafourche and Terrebonne parishes, often in areas located outside of leveed communities.

Fig. 22. Distribution of Minority Populations by Census Block (2010)

Data Source: U.S. Census Bureau

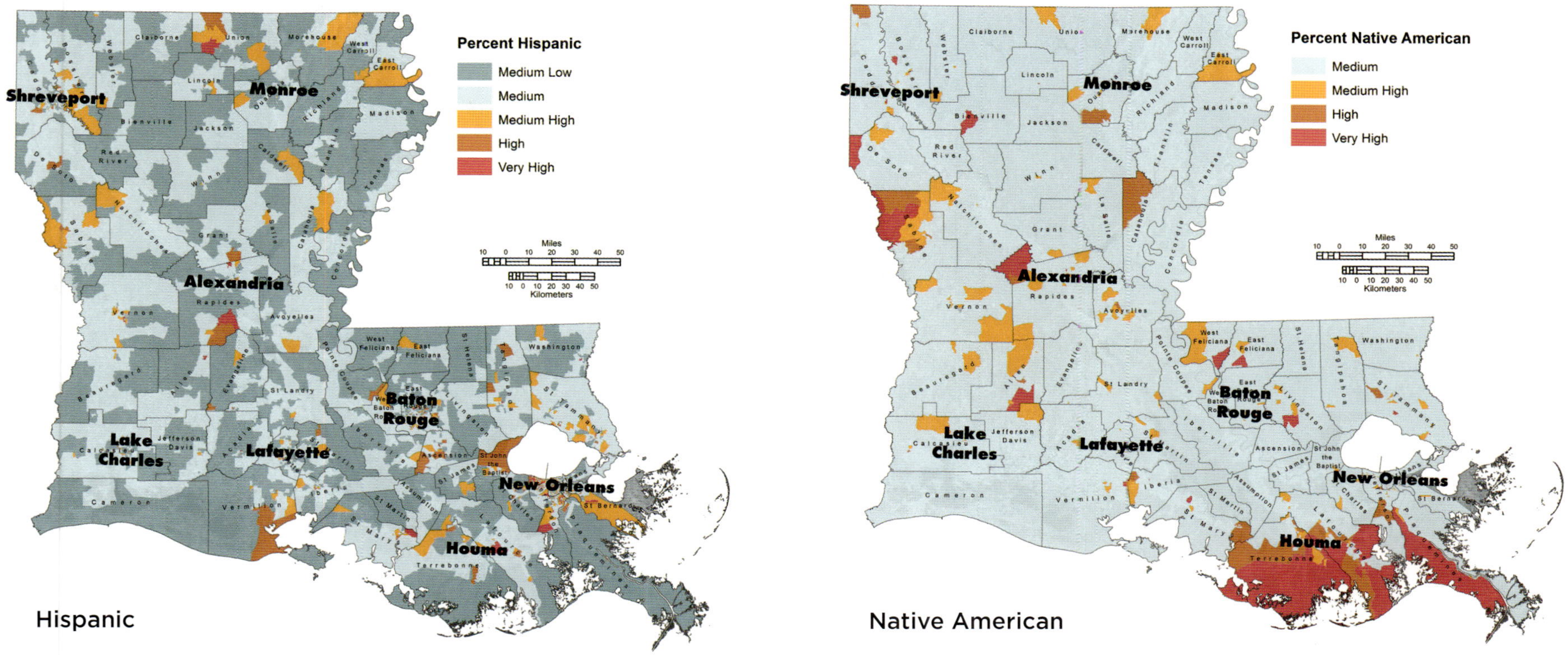

Fig. 23. Elderly Population and Pre-school-Aged Children in Coastal Shoreline and Coastal Watershed Parishes

3 maps

Data Source: U.S. Census Bureau

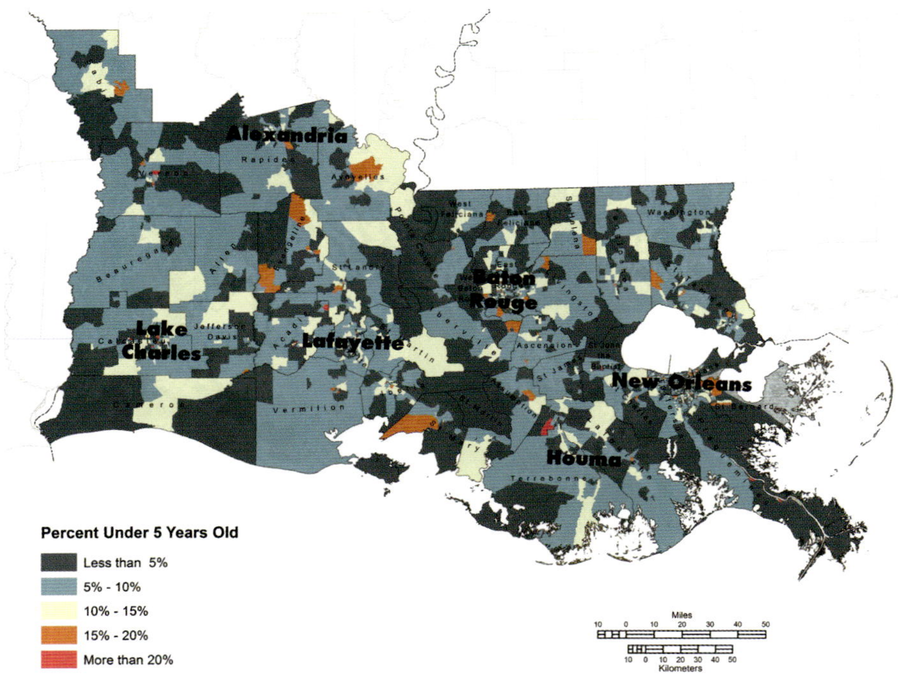

Children

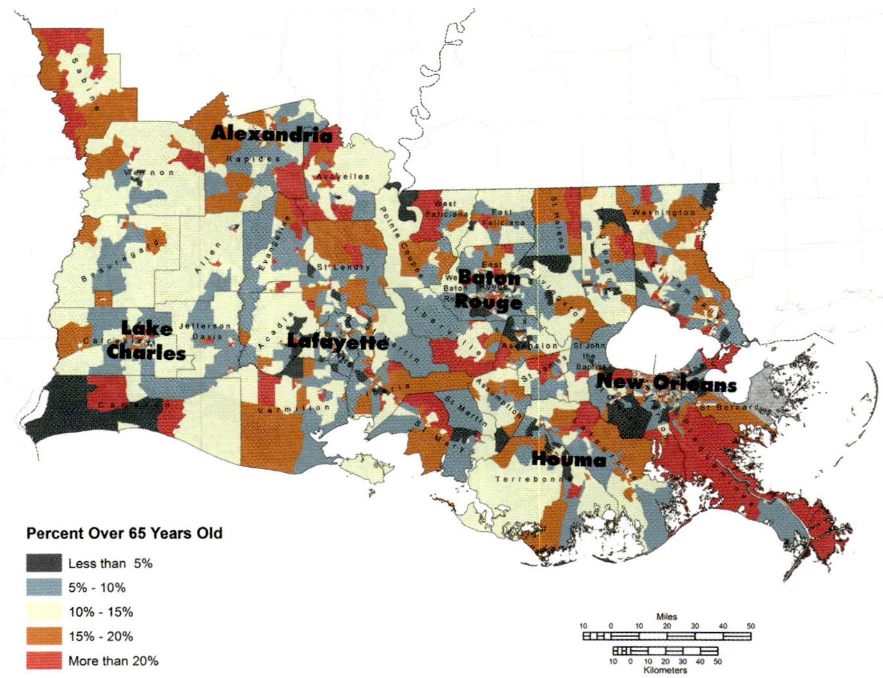

Elderly

Finally, it is important to look at the locations of dependent populations, specifically preschool-aged children and the elderly. These populations are highly dispersed across coastal Louisiana without any well-defined clusters of either group (fig. 23). However, when the two groups are examined together, it is apparent that in most census block groups in the coastal zone, over 20 percent of the population is potentially dependent on caregivers, making them especially vulnerable to coastal hazards.

Fig. 23. Elderly Population and Preschool-Aged Children in Coastal Shoreline and Coastal Watershed Parishes

Data Source: U.S. Census Bureau

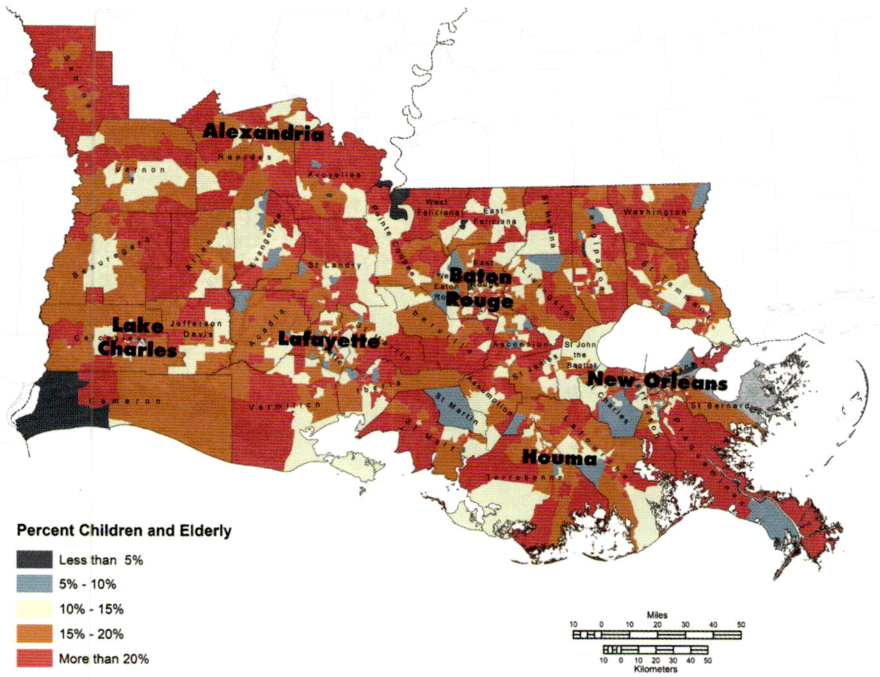

Children/Elderly

CHAPTER 2

CROSSING THE THRESHOLD

The Cascading Effects of Historical Social and Environmental Change

Community resilience provides a different perspective on the relationships that societies have with their environmental setting. Human societies have always strived for stability and attempted to build coping capabilities into their community structures, but the truth is that human societies, by nature, exist in a state of flux, constantly adjusting in subtle ways to any number of internal and external disturbances. The ability to anticipate, respond to, minimize the impacts of, and recover from disruptive events is resilience. Adaptations are the adjustments that society makes to fortify its resilience, to incorporate lessons learned after a calamity. Thus, resilience is a historical process that develops, grows, and perpetuates between disruptive events. It is this historical component of resilience that promotes the development of long-term adaptations to changing environmental conditions, allowing communities to cultivate inherently resilient qualities through time.

Without understanding that resilience is innately a historical process, community residents and policy makers cannot fully and effectively respond to hazard events. Further, communities will be unable to successfully anticipate multi-hazard threats, those that are made more severe by preconditions of social vulnerability and other concurrent stresses or events (Colten et al. 2008). Preexisting, historical trajectories of environmental, political, and economic change ultimately help determine how communities will respond to and recover from hazard events and other social stresses.

A community's response to a disturbance hinges on the type and scale of the disturbance itself. Large-scale environmental disturbances expose populations and social systems to demands and consequences that fall outside the usual realm of human experience. These events can occur suddenly, as in the case of hurricanes, or more insidiously over periods of time that can be measured in decades or even generations, as with coastal land loss and subsidence (Paton 2006). Similarly, innate changes in community structure, such as population growth and decline, migration, deindustrialization, and shifting geographies of poverty, aging, and technological change, operate at multiple geographical and temporal scales. To more effectively understand how human populations respond to these changes, studies in the short term at the community level and in the long term at regional levels are both essential. While short-term, community-scale analyses can reveal bellwether changes and immediate responses to structural or policy changes, longer-term regional-scale analyses will reveal whether and why intrinsic changes to community structure resonate enough to promote broader change and enhanced resilience (Pendall et al. 2007).

At the regional level, population shifts from cities to rural areas, and vice versa, are generally considered to be an effect of economic changes, with cities experiencing a loss of employment opportunities and a resultant outmigration of the population (Pallagst et al. 2013). Indeed, counter-urbanization, an extension of both deindustrialization and suburbanization with populations shifting from large industrial metropolitan areas to small rural communities, is often viewed as contributing to a loss of community resilience. In coastal Louisiana, these processes are further confounded by increasing levels of risk faced by residents of many small rural communities, in effect driving them farther inland into more protected urbanized areas. It is this tension between social and environmental influences played out on the landscape that underlies the resilience of coastal communities.

This situation is further complicated when matching geographic and time scales. A decade might provide a sufficient time horizon to understand a region's resilience in the face of a single shock, be it an extreme hazard event or some type of policy adjustment; however, it might take a generation to learn how a region is changing in the face of repeated shocks. On the other hand, understanding the resilience of a region in the face of slow-moving external challenges, such as coastal land loss, might require observation over multiple generations (Pendall et al. 2007). Historical responses to past events ultimately shift the degree of inherent resilience possessed by communities, sometimes noticeably and sometime subtly. In many cases, momentum from the past is good for maintaining landscapes and community. It creates value, generates wealth, and maintains a distinct sense of community and place deeply rooted in the cultural heritage of the people. But occasionally that momentum leads a community down a path toward environmental unsustainability (Campanella 2010). In such cases, historical momentum, most often driven by the community's desire to resurrect pre-disaster patterns of culture and human interaction, virtually guarantees that communities restore their pre-disaster physical character (Mileti and Passerini 1996), which can reduce levels of inherent resilience. The post-disaster situation, whether it reflects the direct consequences of the disruption itself or the effectiveness of the recovery and rebuilding process, will present community members with a new reality that may differ in fundamental ways from conditions before the transformative event. It is the changed physical, social, and psychological reality that people must adapt to (Paton 2006). A previously resilient community may suddenly find itself unable to fully recover from an event, regardless of the scale of that event. While the population may return, the community may, in fact, be weakened and less resilient.

Population recovery and return to prior functional levels after a major event may apply well to dramatic events with clearly identifiable start and end points but not to ongoing, chronic stress without a clear beginning or end (Schetter and Dolbier 2011). When examined in combination with the ongoing effects of climate change, coastal land loss, and other environmental and ecological stressors, the impacts of acute, episodic events serve as a source of ongoing, chronic stress to coastal communities. In this environment, the return of population levels to those that existed prior to a singular acute storm event, for example, should not be taken as the sole indicator of community resilience. By examining chronic environmental stress along with the ongoing stress associated with low socioeconomic status, for example, policy makers are better able to determine the impacts of acute, singular hazard events or disturbances on community resilience.

NATURAL AND INDUSTRIAL HAZARDS

Flooding is the most commonly occurring natural hazard in coastal Louisiana. Flooding is generally defined as the inundation of areas not commonly submerged. This inundation can be the result of meteorological events or the overbanking of rivers and streams. The most dramatic flooding often accompanies large-scale hazard events such as hurricanes and storm surge.

Fig. 24. Historical Density of Extreme Tropical Weather Events in Louisiana (1940–2010)

7 maps

Map surface was adapted from hurricane track data weighted by maximum wind speed and adjusted to account for potential impact zones.

Data Source: NOAA, National Weather Service, National Hurricane Center

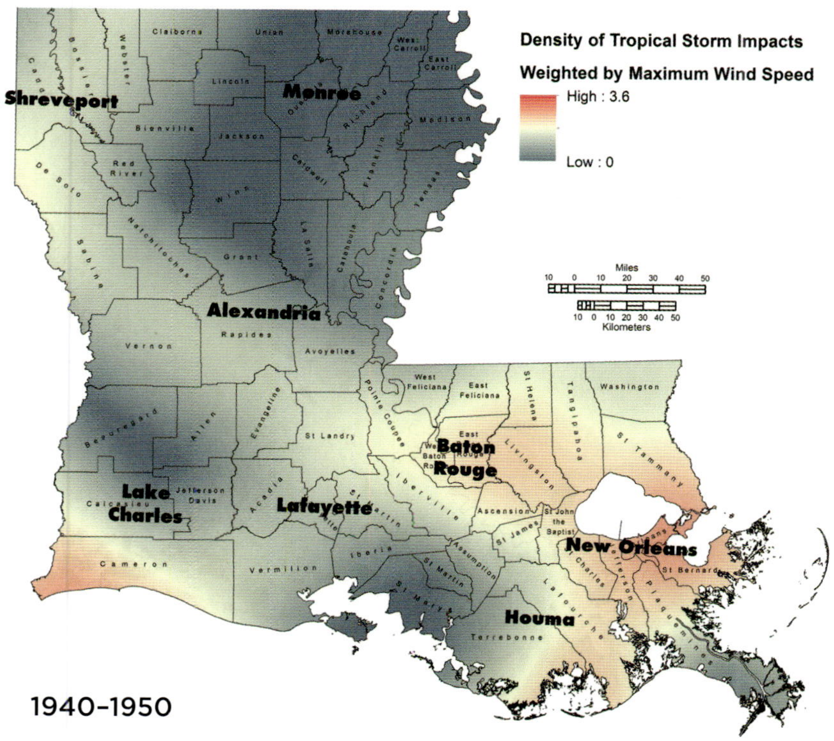

1940–1950

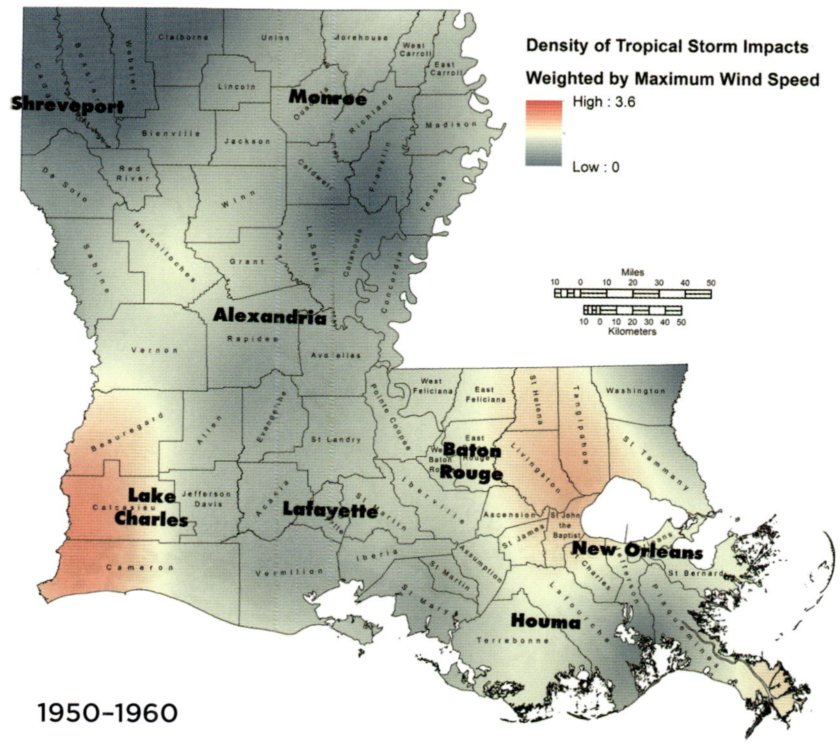

1950–1960

Louisiana has long been the epicenter of tropical storm and hurricane activity in the Gulf of Mexico. Hurricanes are the most extreme tropical weather events occurring in the region, accompanied by devastating winds of 74 to 160 miles per hour and storm surges often in excess of 20 feet in depth. The full impacts of hurricanes in Louisiana are felt far from the center of the storm as hurricane-strength winds often extend 50 miles or more from the epicenter. Over time, most of Louisiana's Gulf Coast has experienced a number of extreme tropical weather events and the associated risks and hazards (fig. 24). However, there has been a significant amount of temporal and spatial variability in the density of storm strikes across Louisiana, with impact zones dispersed widely across the width of the state. The first decade of the twenty-first century has seen a sharp uptick in the density of these events.

Fig. 24. Historical Density of Extreme Tropical Weather Events in Louisiana (1940–2010)

Data Source: NOAA, National Weather Service, National Hurricane Center

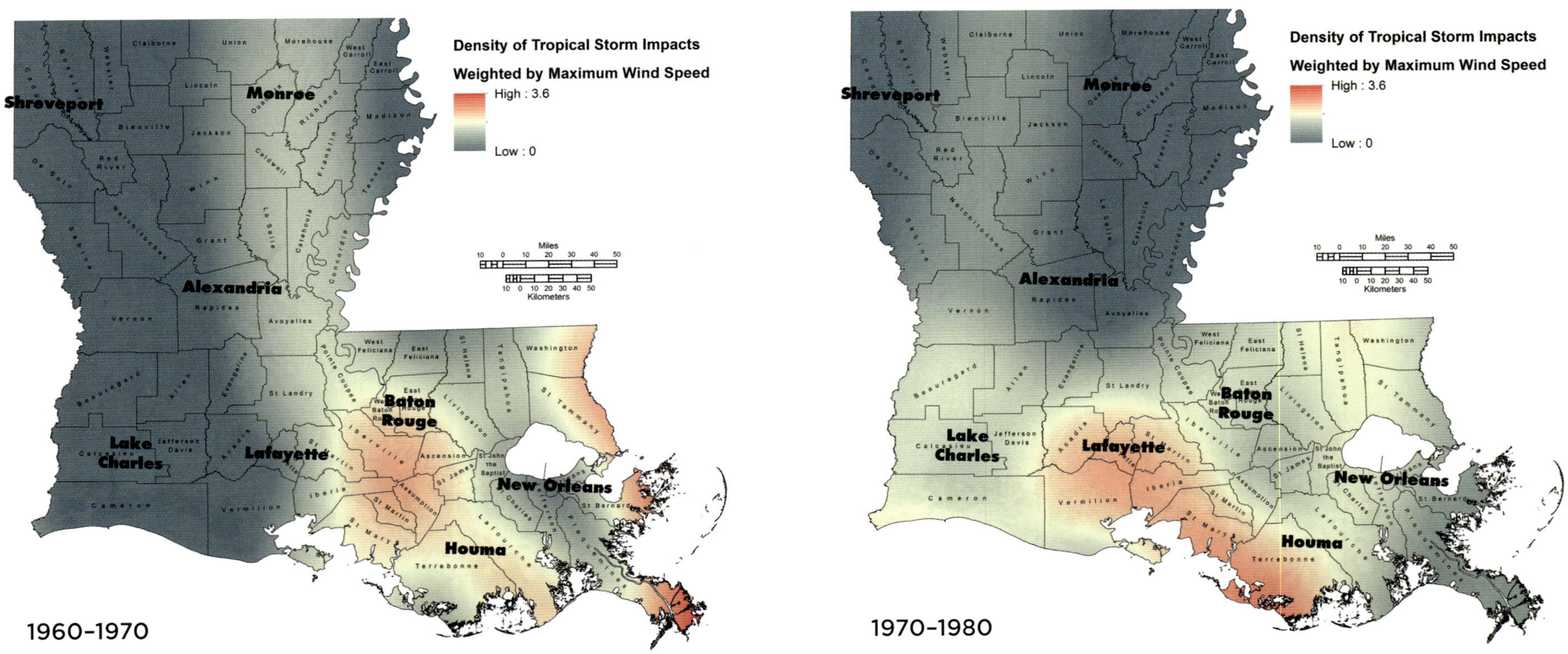

Fig. 24. Historical Density of Extreme Tropical Weather Events in Louisiana (1940–2010)

Data Source: NOAA, National Weather Service, National Hurricane Center

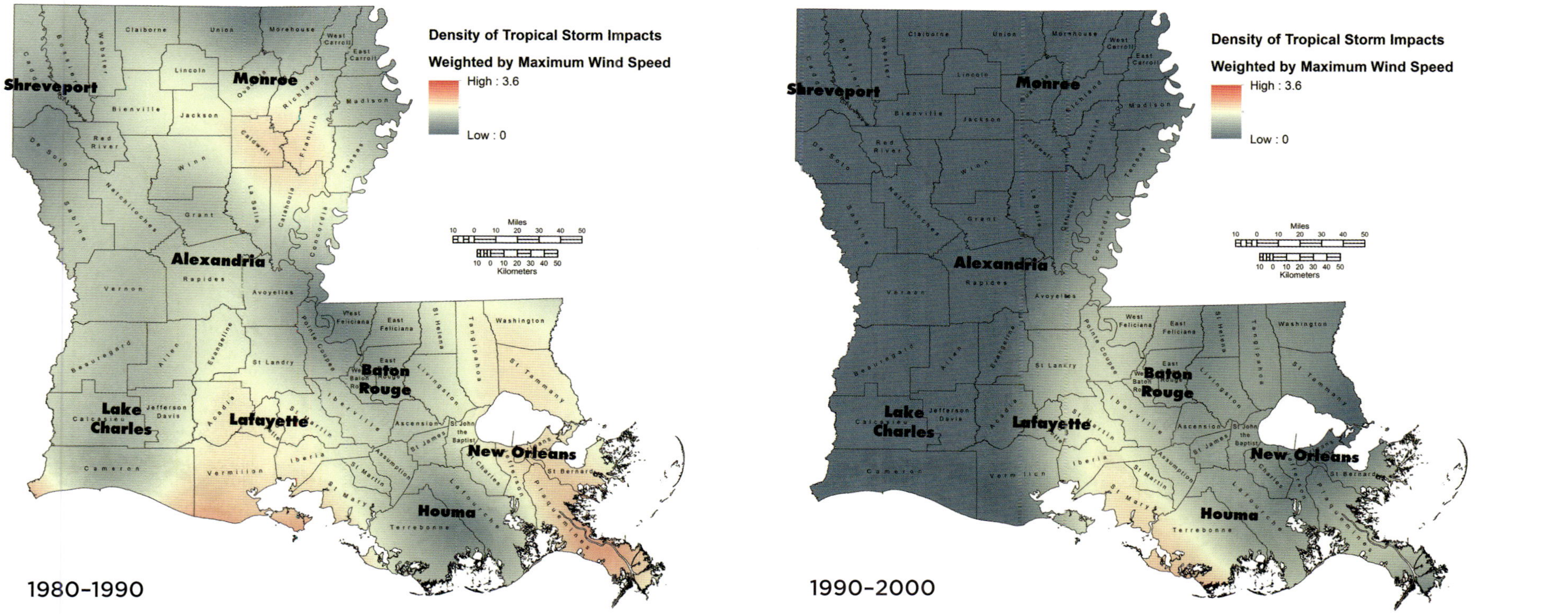

Fig. 24. **Historical Density of Extreme Tropical Weather Events in Louisiana (1940–2010)**

Data Source: NOAA, National Weather Service, National Hurricane Center

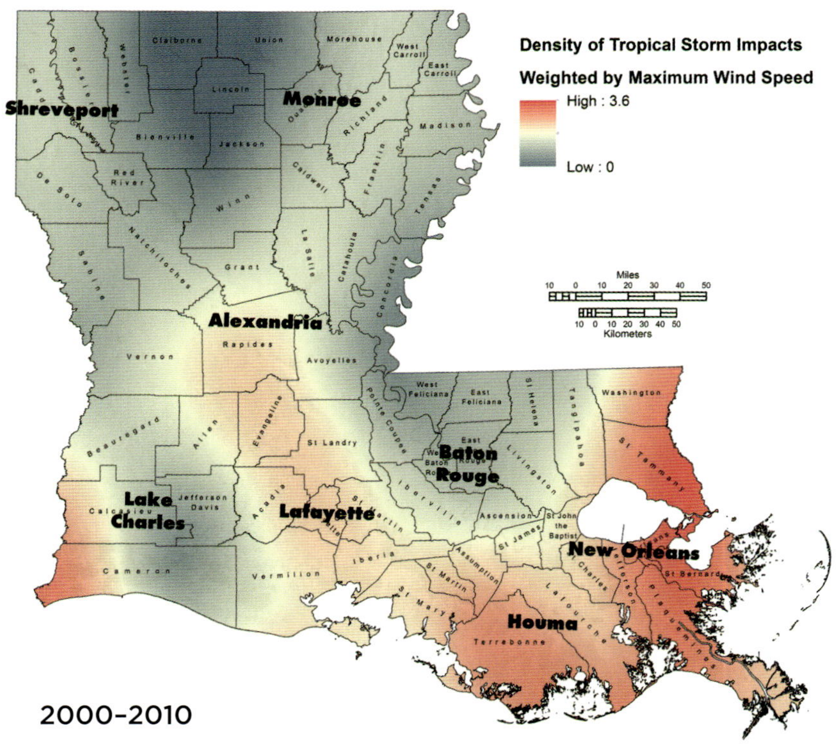

2000–2010

Fig. 25. Total Days above Flood Stage at Selected River Gauges (1950–1960)

Data Source: U.S. Department of Commerce, Weather Bureau

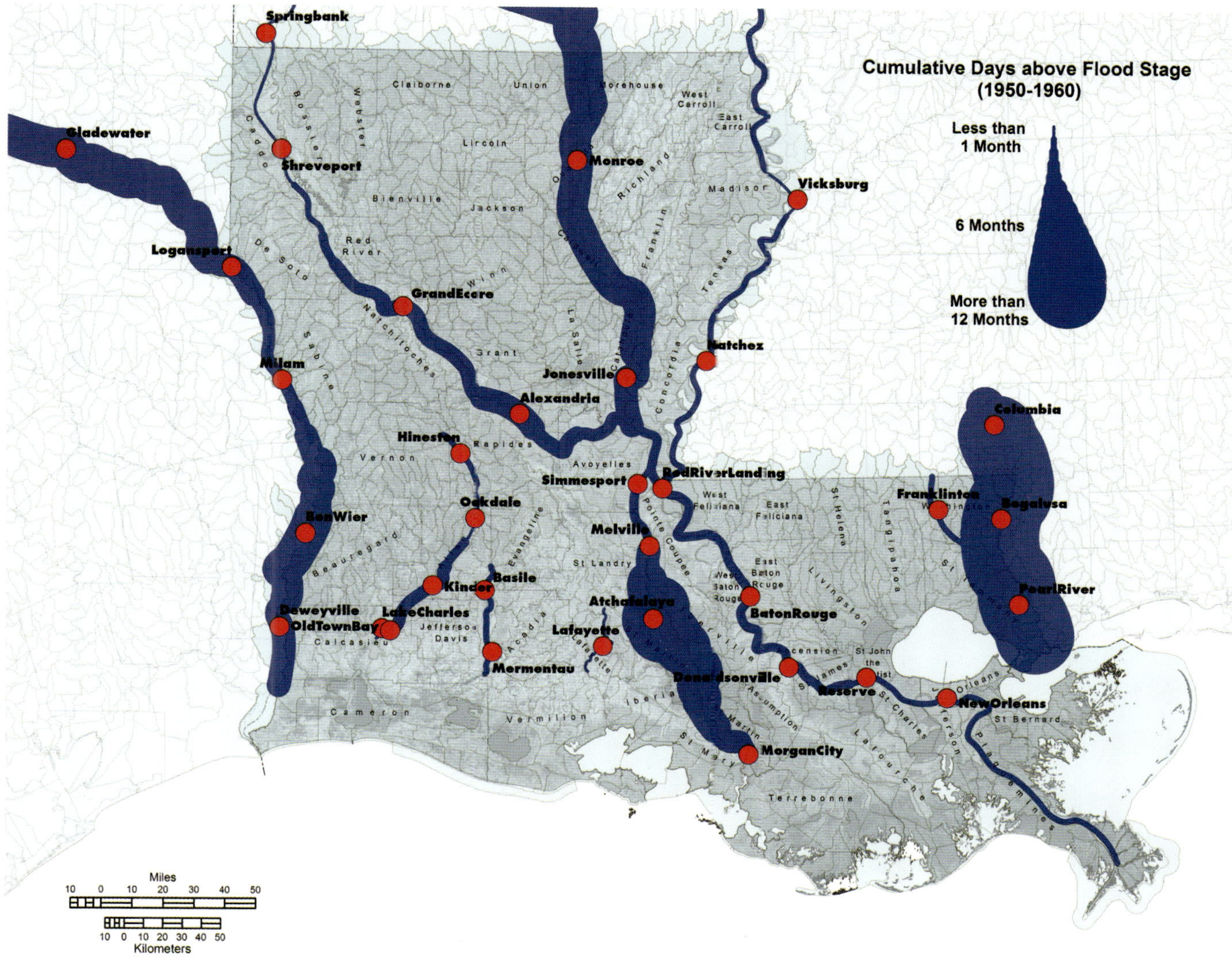

Riverine flooding is another ever-present threat to residents of Louisiana (fig. 25). Generally more predictable than tropical weather events, annual river flooding results from a combination of precipitation and runoff into the river upstream. During the spring snow thaws in the northern United States, for example, massive volumes of water drain into rivers and waterways, many of which ultimately flow through Louisiana and into the Gulf of Mexico. If the river exceeds its carrying capacity, it will overrun and floodwaters will spill into the surrounding floodplain.

Fig. 26. Total Days above Flood Stage at Selected River Gauges (2000–2010)

4 maps

Data Source: U.S. Department of Commerce, Weather Bureau

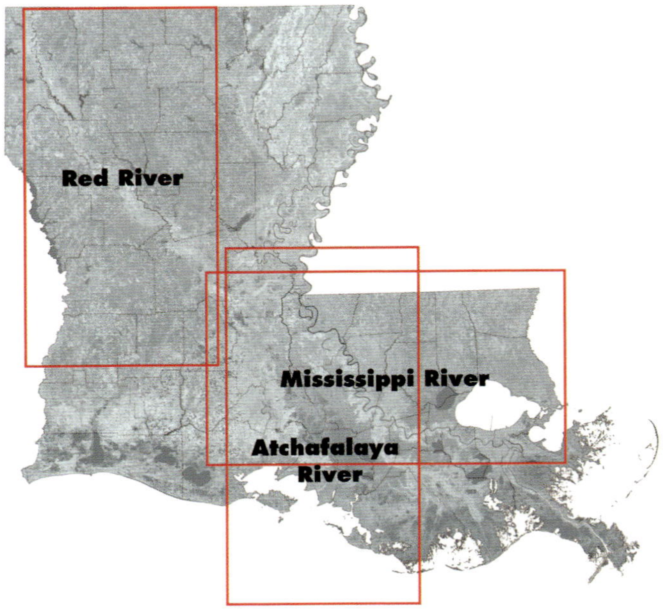

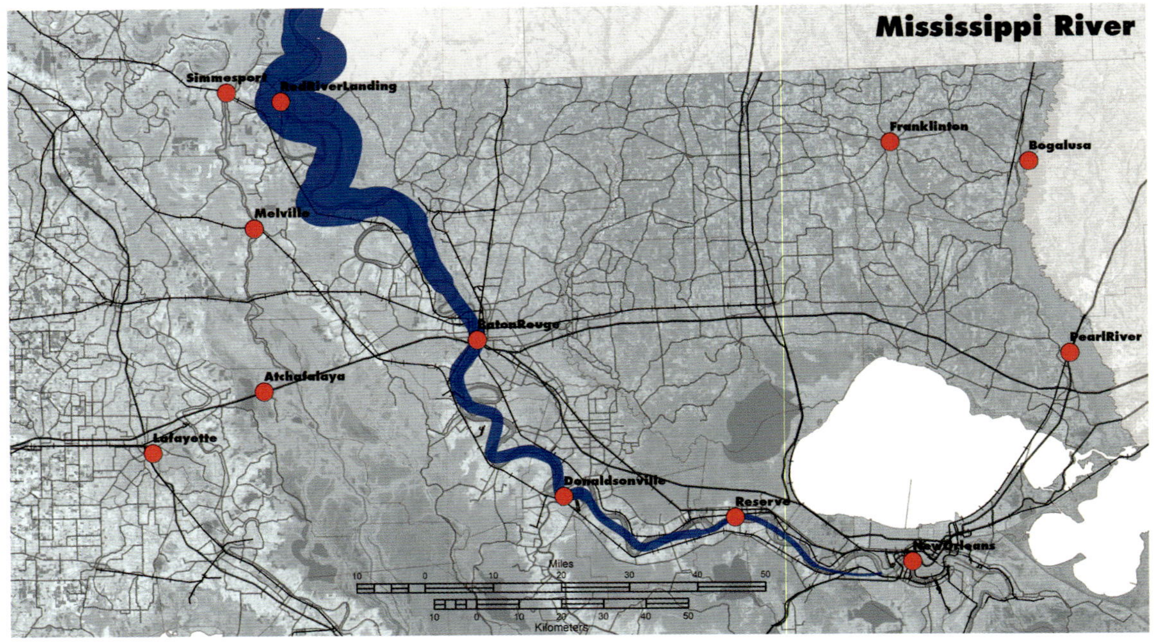

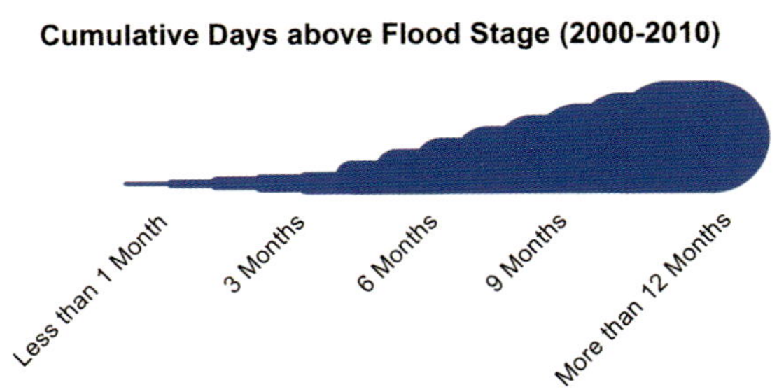

The effects of river control structures on stage and flood events in specific river basins are notable (fig. 26). The reduction of flood events in the lower Red River, for example, coincides largely with engineering actions to realign and stabilize the river banks and the construction of a series of navigation locks and dams. Other river control projects are designed to remove flow from one system and divert it into another. The Old River Control Structure, located at the juncture of the Mississippi River, the Red River, and the Atchafalaya River, for example, is designed to reduce flooding in the lower Mississippi River by diverting the high water flows into the Atchafalaya River.

Fig. 26. Total Days above Flood Stage at Selected River Gauges (2000–2010)

Data Source: U.S. Department of Commerce, Weather Bureau

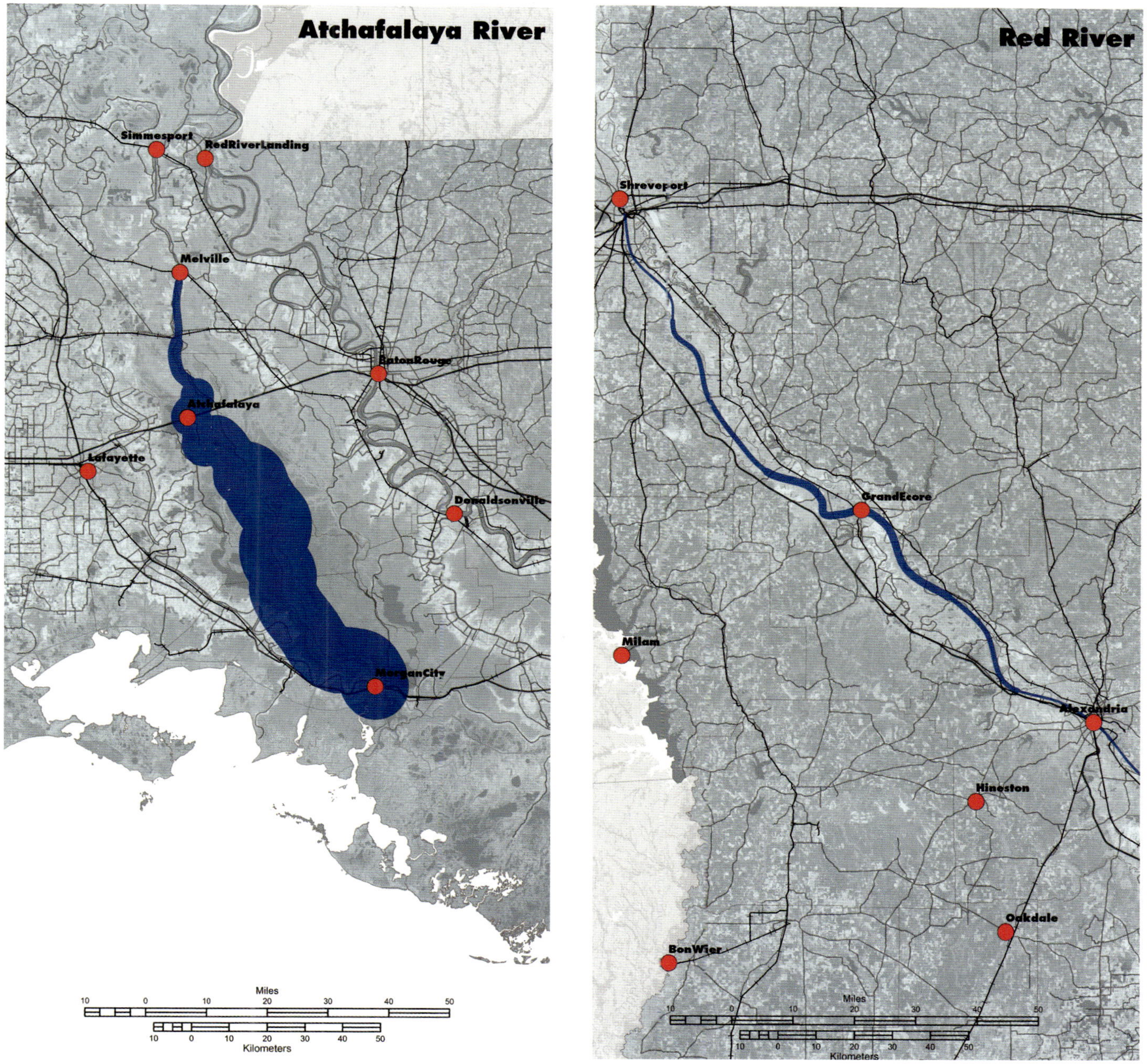

Fig. 27. Total Number of Extreme Rainfall Events (1940–2010)

Map surface was adapted and interpolated from individual weather station data.

7 maps

Data Source: NOAA Regional Climate Centers, Applied Climate Information System

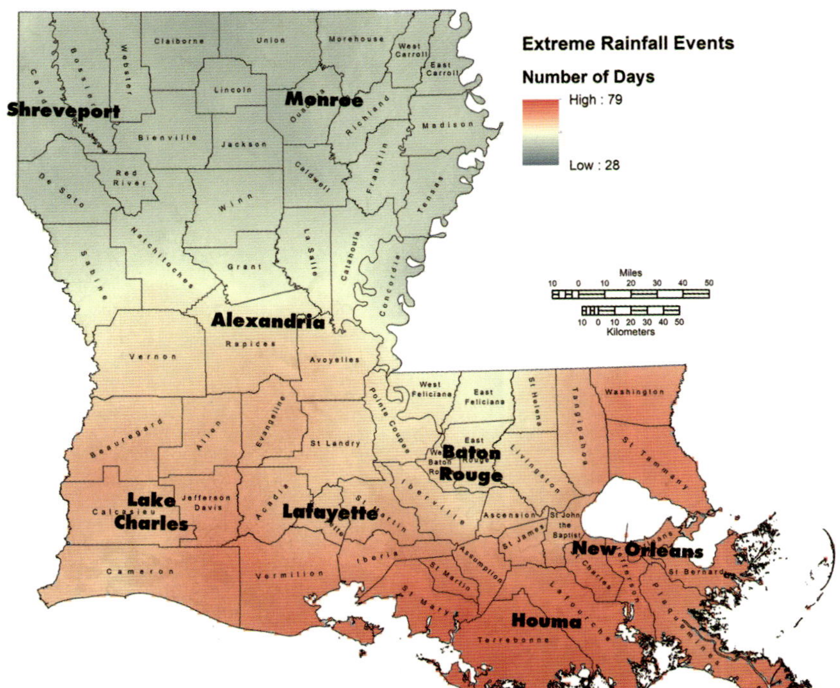

1940–1950

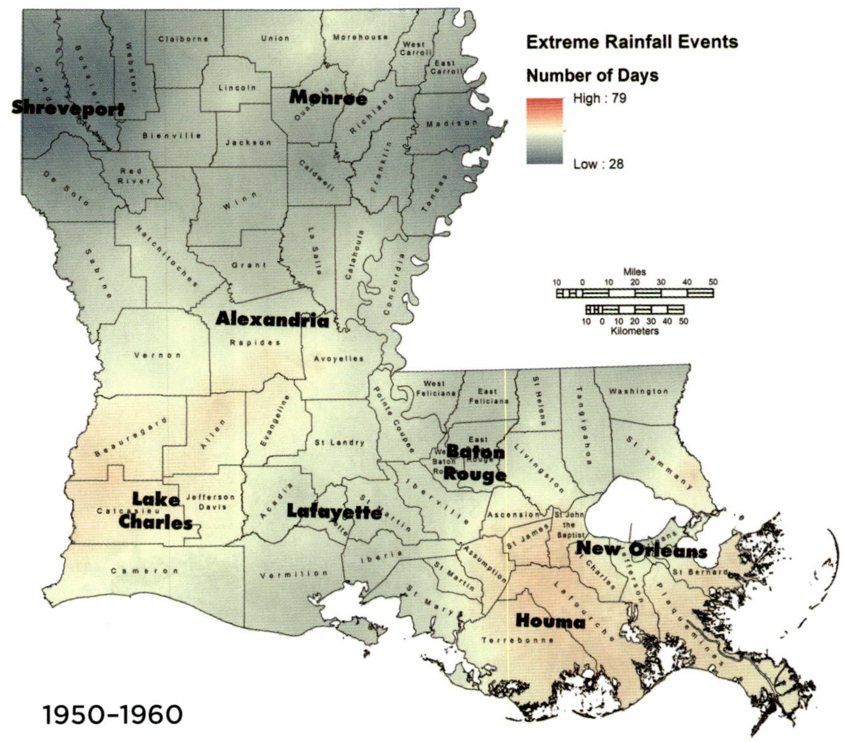

1950–1960

The timing and spatial distribution of precipitation present another flood risk to communities in Louisiana's coastal zone (fig. 27). Extreme precipitation events have the potential to induce flooding across a wide range of stream basins. Decadal precipitation totals highlight the cyclical nature of such extreme events, cycles driven in large part by the El Niño/Southern Oscillation, or ENSO, an irregularly timed pattern of changing weather regimes impacting large portions of the United States, including Louisiana. ENSO events recur every two to six years and bring in higher precipitation levels, enhancing the risk of coastal flooding during ENSO years.

Fig. 27. Total Number of Extreme Rainfall Events (1940–2010)

Map surface was adapted and interpolated from individual weather station data.

Data Source: NOAA Regional Climate Centers, Applied Climate Information System

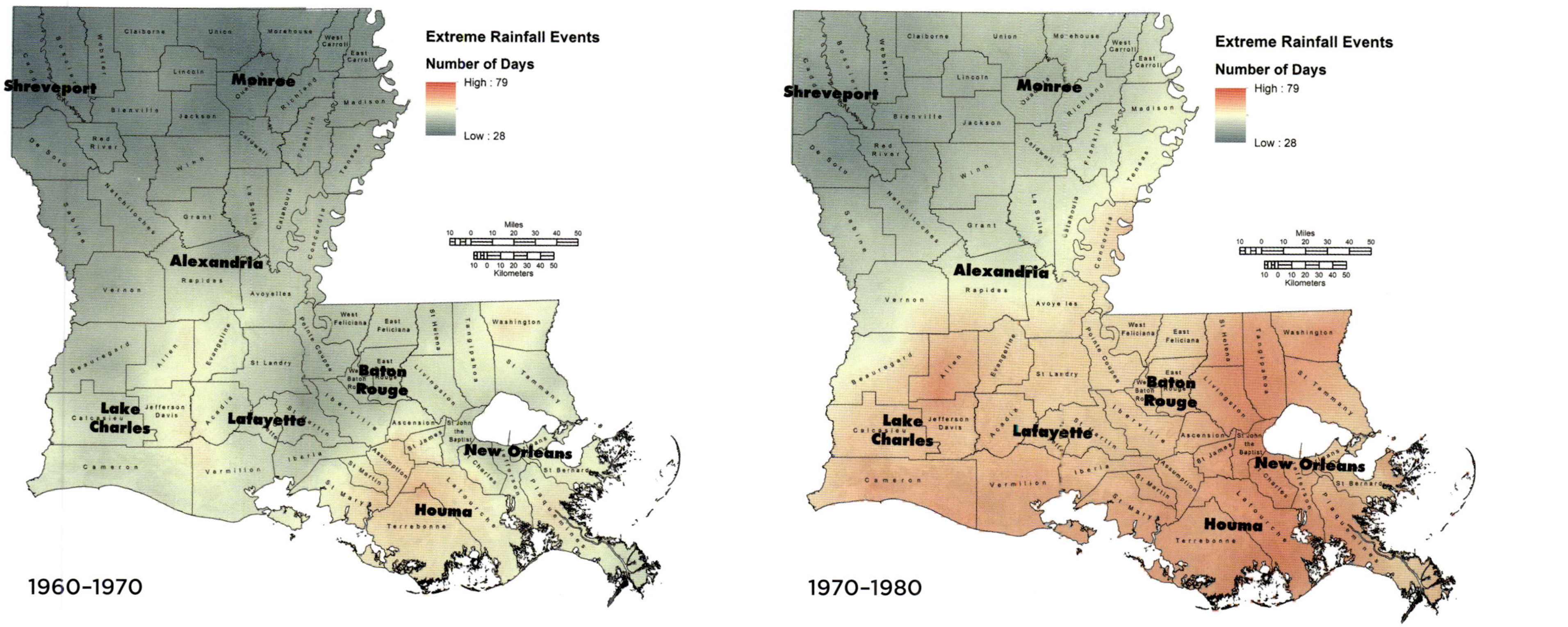

Fig. 27. Total Number of Extreme Rainfall Events (1940–2010)

Map surface was adapted and interpolated from individual weather station data.

Data Source: NOAA Regional Climate Centers, Applied Climate Information System

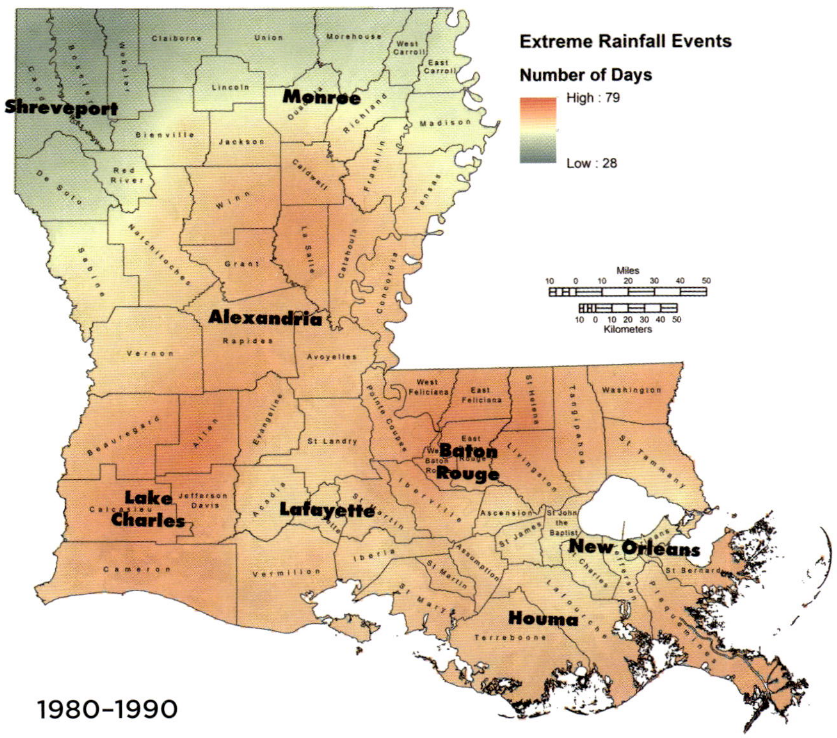

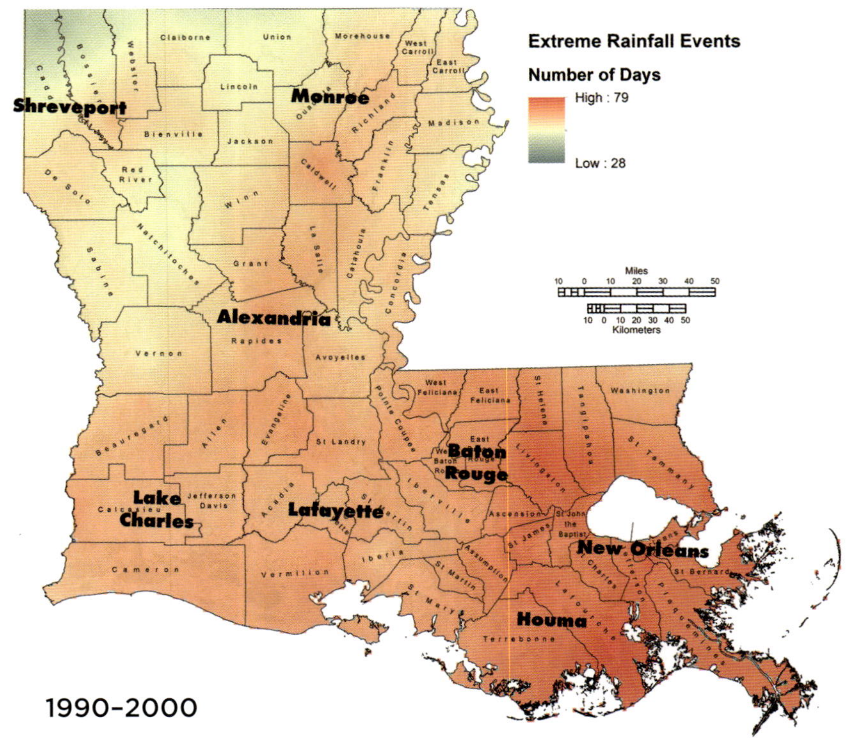

Fig. 27. Total Number of Extreme Rainfall Events (1940–2010)

Map surface was adapted and interpolated from individual weather station data.

Data Source: NOAA Regional Climate Centers, Applied Climate Information System

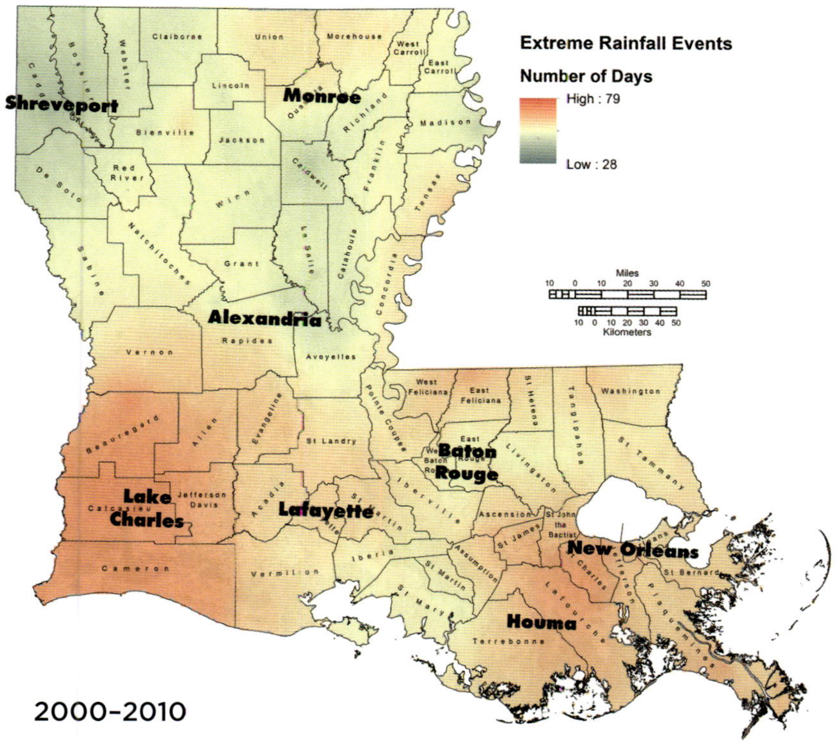

2000–2010

Fig. 28. Total Number of Days in Moderate to Severe Drought (1940–2010)

Map surface was adapted and interpolated from individual weather station data.

7 maps

Data Source: NOAA Regional Climate Centers, Applied Climate Information System

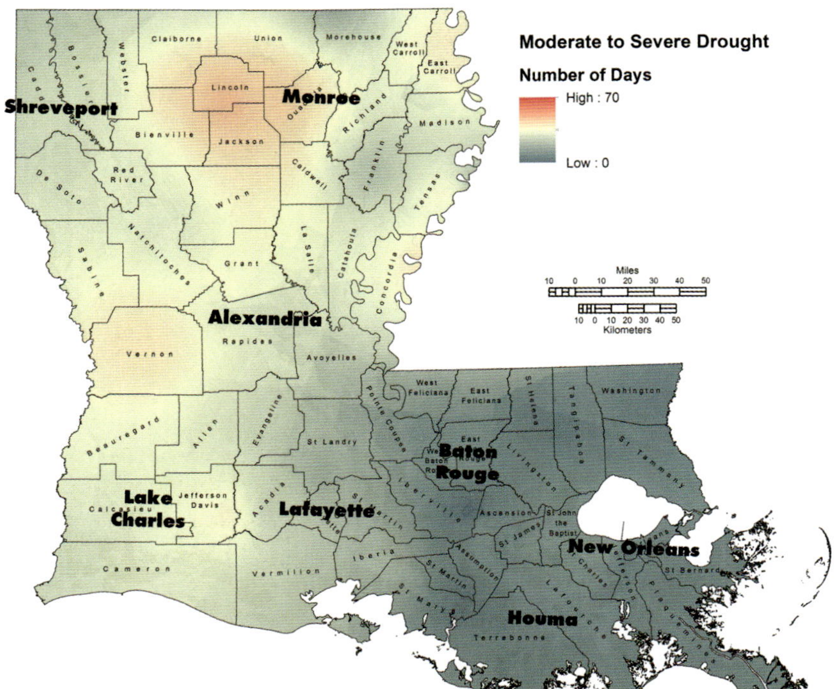

1940–1950

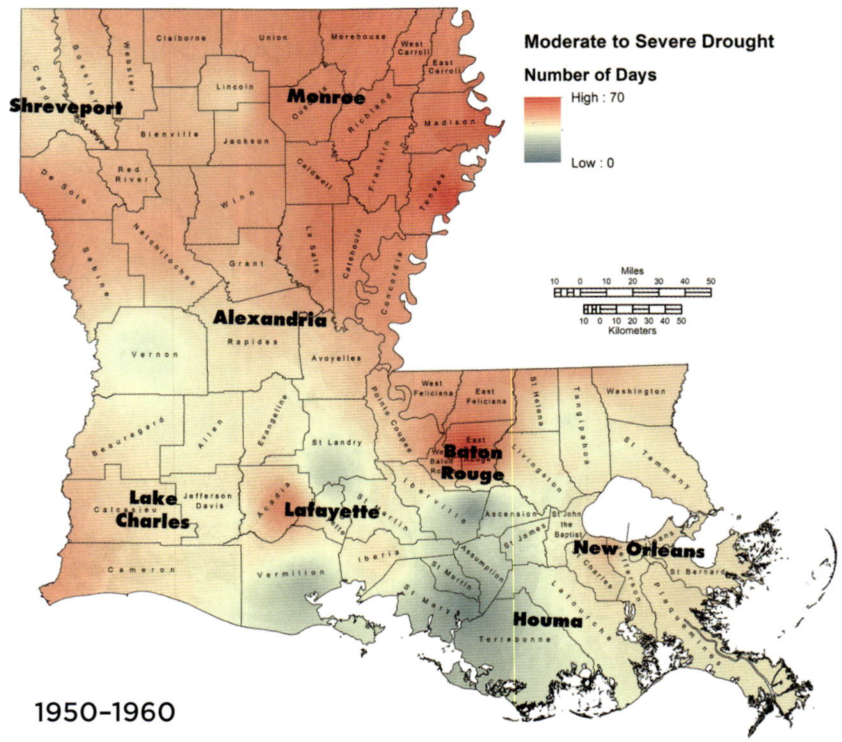

1950–1960

While events such as coastal flooding and sea level rise tend to dominate much of the hazards discussion in Louisiana, in some portions of the state, moderate to severe drought conditions represent an equally pressing matter. La Niña periods often result in long periods of excessive dryness and low soil moisture, culminating in excessive drought conditions, concerns for both the agriculture and forestry industries in Louisiana (fig. 28). In some cases, the state has even experienced both flood and drought conditions in the same location, due to the fact that much of the flooding originates from rivers or the Gulf of Mexico while drought conditions are driven largely by rainfall events.

Fig. 28. Total Number of Days in Moderate to Severe Drought (1940–2010)

Data Source: NOAA Regional Climate Centers, Applied Climate Information System

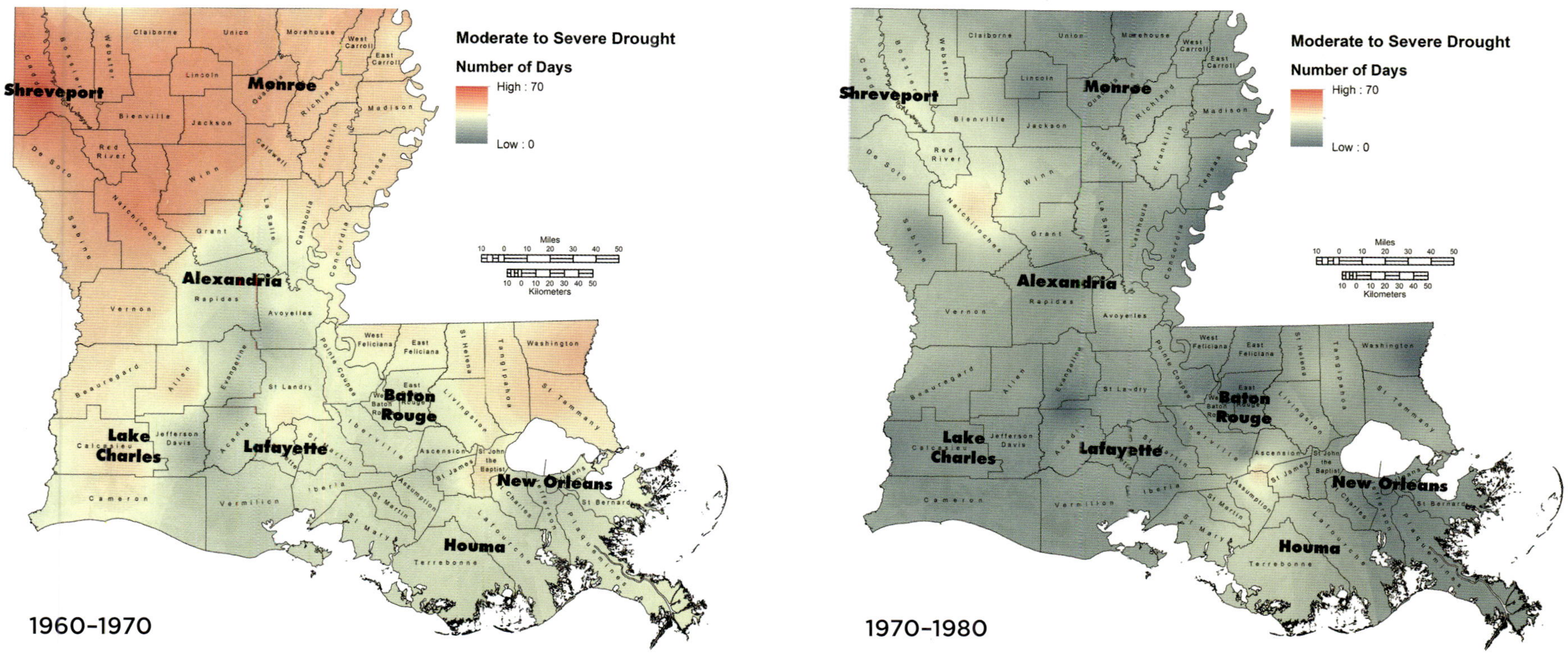

Fig. 28. Total Number of Days in Moderate to Severe Drought (1940–2010)

Data Source: NOAA Regional Climate Centers, Applied Climate Information System

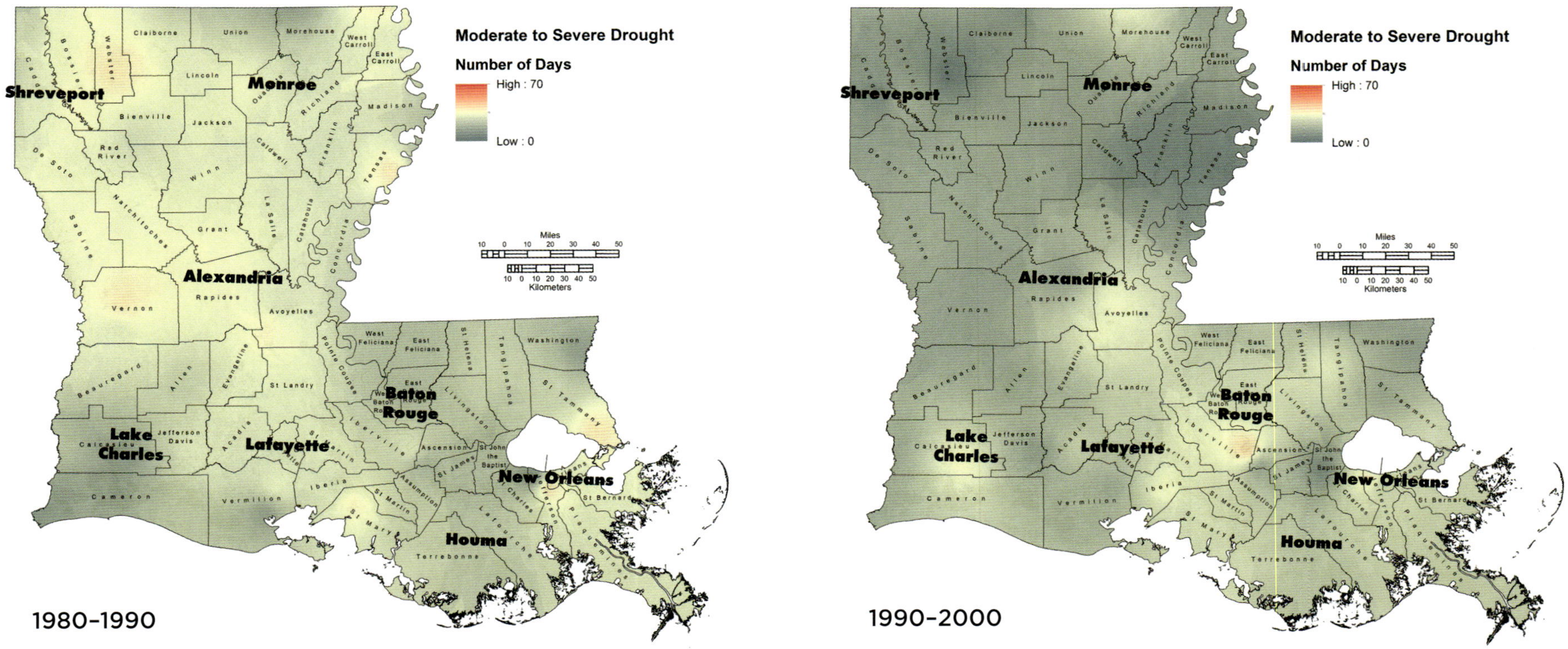

Fig. 28. **Total Number of Days in Moderate to Severe Drought (1940–2010)**

Data Source: NOAA Regional Climate Centers, Applied Climate Information System

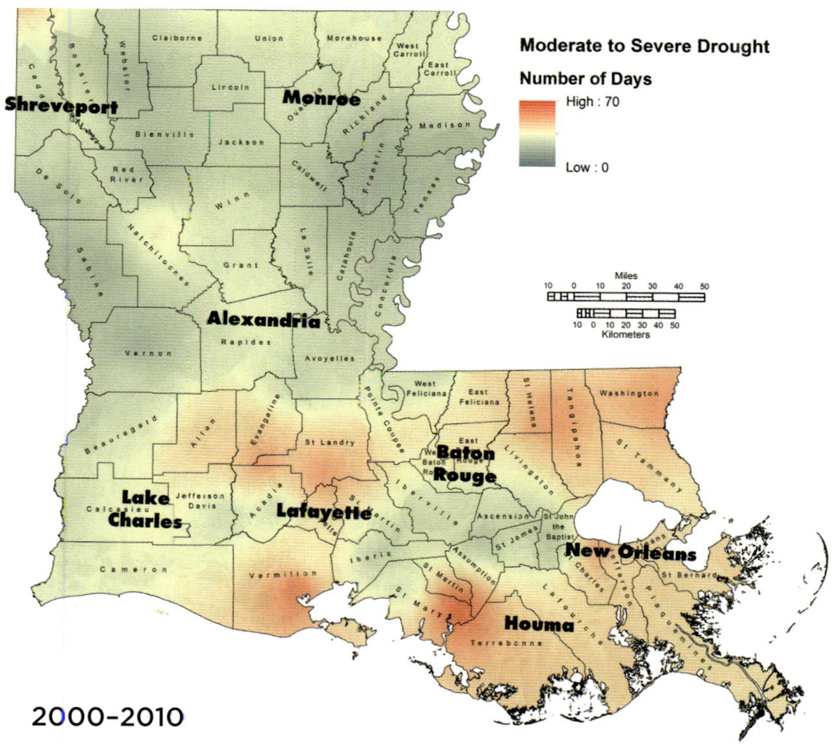

2000–2010

Fig. 29. Land Area Change by Census Block

4 maps

Data Sources: Couvillion et al. 2011; U.S. Census Bureau

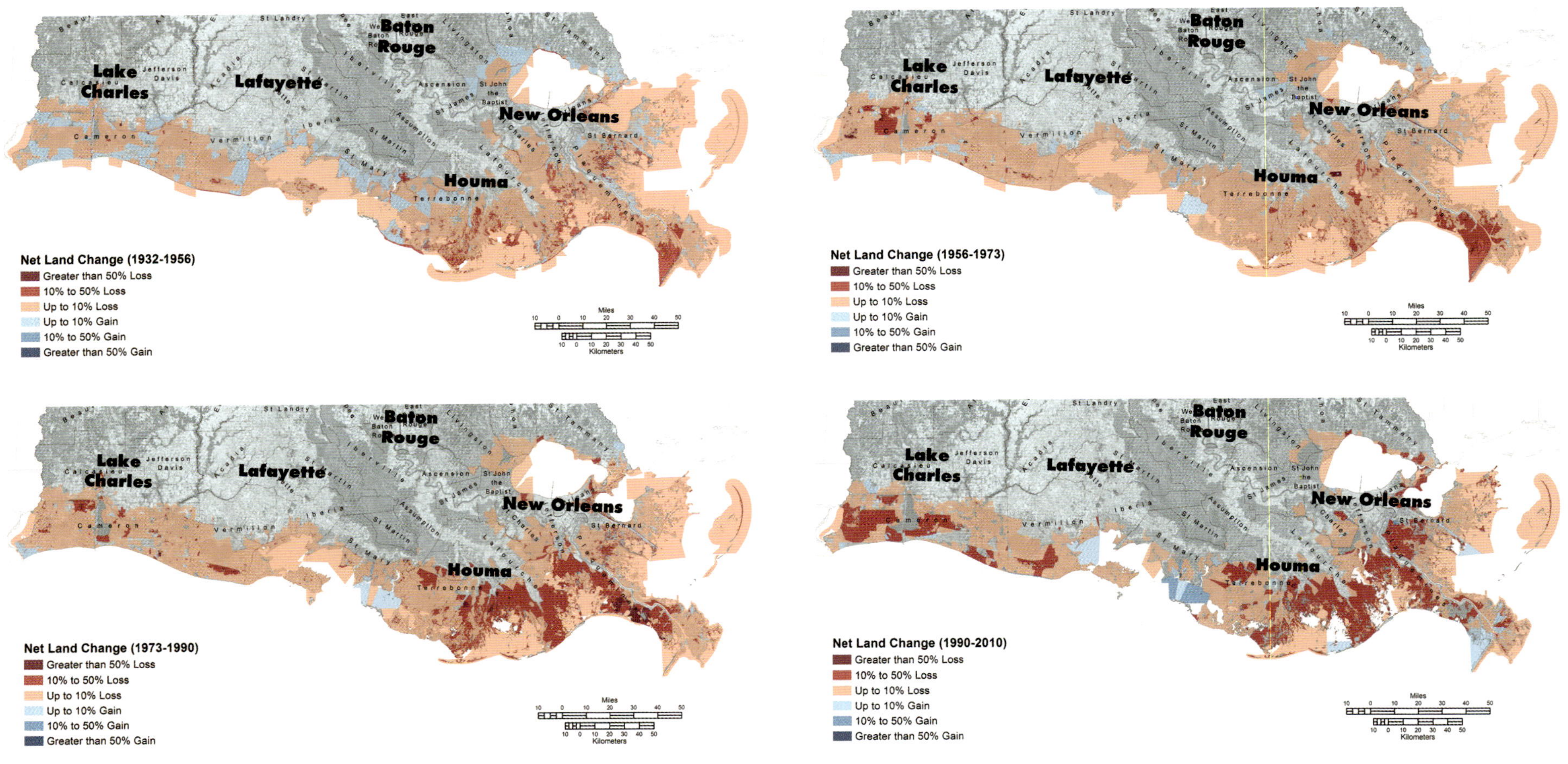

The cumulative impact of changing environmental conditions over time has been devastating to the physical geography and ecology of coastal Louisiana. Coastal erosion, due to a combination of human-induced and natural conditions, and geophysical processes such as coastal subsidence resulted in substantial land loss in coastal Louisiana throughout the twentieth century (fig. 29). This process intensified over the last quarter of the twentieth century and into the twenty-first. In fact, since 1990, several census blocks in the coastal zone have lost more than 50 percent of their land area. Most of these blocks are south of some of the larger cities in Louisiana's coastal zone, including Lake Charles, Houma, and New Orleans. While coastal land loss does not have the immediate impacts on coastal communities that pulse events such as storms and river floods have, it is a more insidious threat, slowly changing the physical geography and ecology of the entire coast that these communities have come to depend on for their livelihoods and well-being.

Fig. 30. Active Toxic Release Inventory Sites in Louisiana and Selected Communities (2010)

Data Source: U.S. Environmental Protection Agency, Toxic Release Inventory

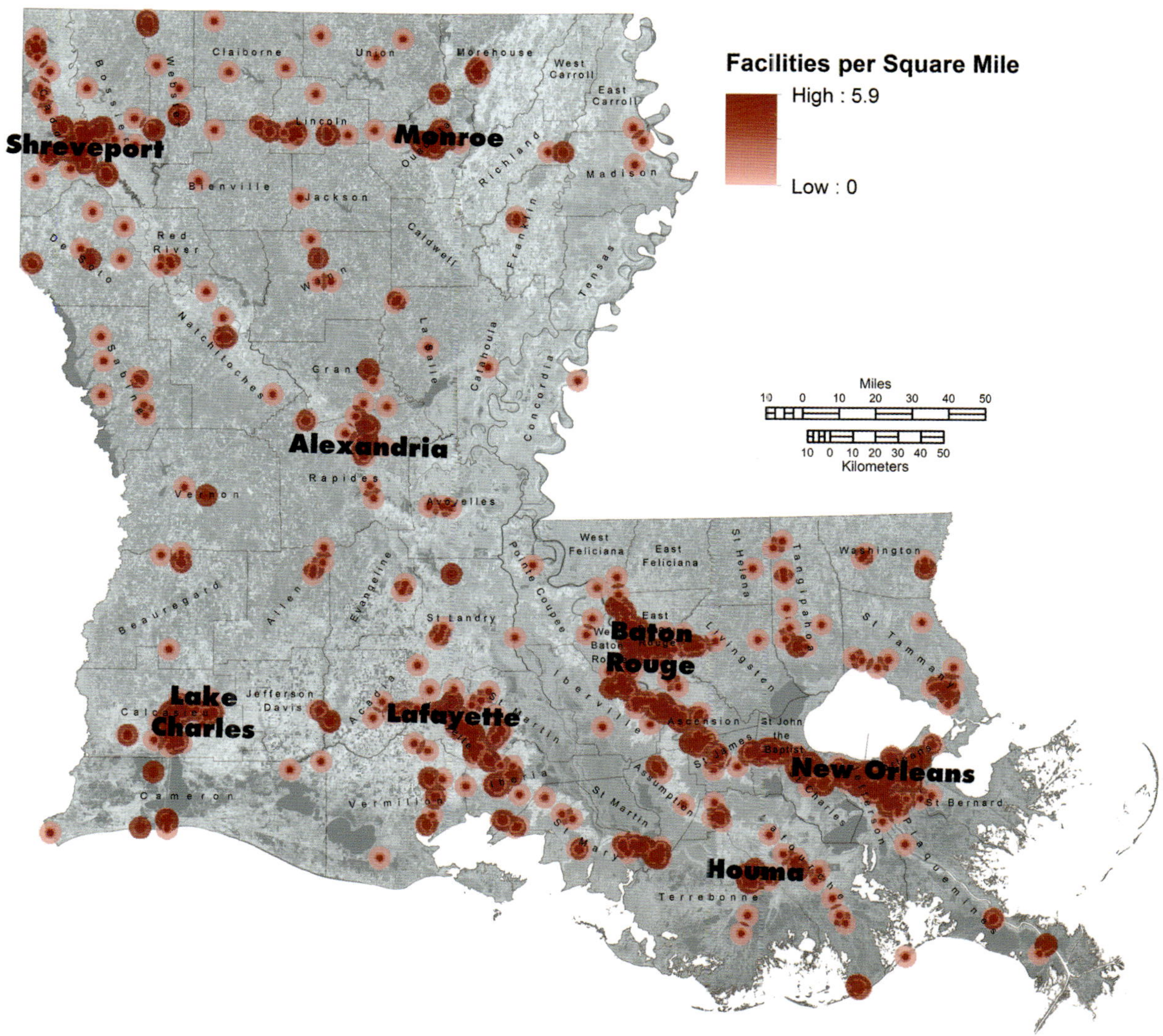

The natural threats to coastal communities, whether enhanced by human intervention or otherwise, represent one category of risk inherent to life in coastal Louisiana. Industrial threats are another omnipresent hazard that coastal residents must face. The state's waters that are the very lifeblood of many coastal residents are also valuable resources for Louisiana's industrial facilities (fig. 30). Extracting and processing the large reserves of oil and natural gas in the coastal zone requires a substantial base of industrial infrastructure, which in turn requires large amounts of water. Petrochemical processing, which occurs upstream, also relies on ready access to the state's waters. This petroleum and petrochemical industrial complex provides tremendous economic advantages to Louisiana but is also a source of environmental and ecological risk to residents and ecosystems located nearby.

Fig. 31. Historical Oil Spills in Coastal Louisiana and Selected Communities (1970–2014)

Data Source: U.S. Coast Guard, National Response Center

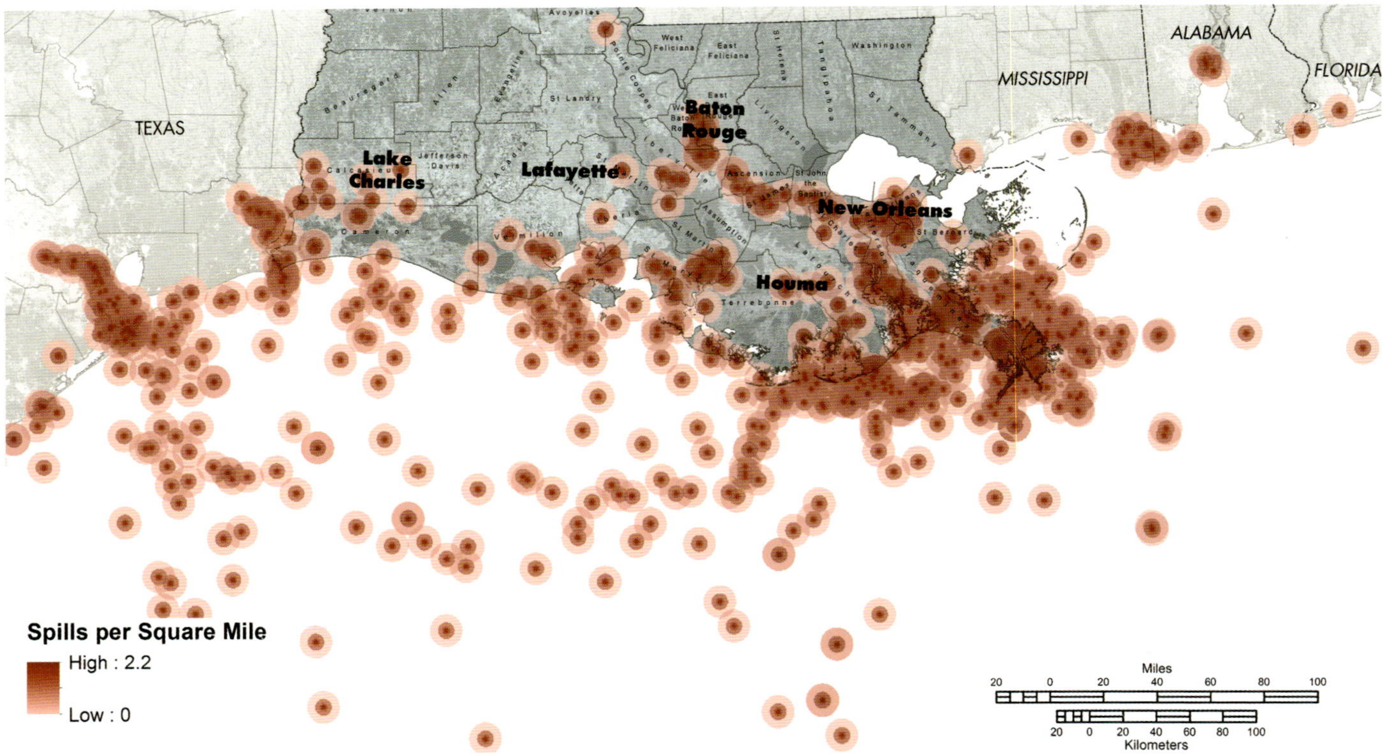

Hazardous facilities are one source of industrial risk to coastal communities. Oil spills are another. Some spills occur at extraction sites, whether on- or offshore (fig. 31). Many more spills occur in the pipelines that transport the product, particularly where they cross roadways, rivers, or bayous. While some of these spills clearly track with the locations of inland petroleum and petrochemical processing plants, many more occur offshore. The U.S. Coast Guard National Response Center is the federal agency that receives all reports of releases involving hazardous substances and oil. The center's data reveal that the vast majority of reported oil spills in the Gulf of Mexico occur in southeast Louisiana. Inland oil spills center on New Orleans, Baton Rouge, and the stretch of the Mississippi River between these cities.

CHANGES IN NATURAL RESOURCE EXPLOITATION

Natural and industrial hazard events that reduce either the supply of or the demand for natural resources have a direct bearing on the economic health, vitality, and resilience of coastal communities. An example of a natural hazard would be a tropical weather event, whereas industrial hazards include oil spills and other hazardous materials releases. Other hazards may in fact be planned, such as freshwater diversion projects designed to rebuild coastal wetlands by introducing sediment-laden river water into brackish or saline environments. In this case, the shift or loss of valuable habitat must be weighed against the long-term benefits of rebuilding land that could ultimately provide protection to a great many coastal residents.

Fig. 32. Pounds of Fish and Shellfish Landed at Major Ports

6 maps

Data Sources: U.S. Department of the Interior, Bureau of Commercial Fisheries; NOAA, National Marine Fisheries Service

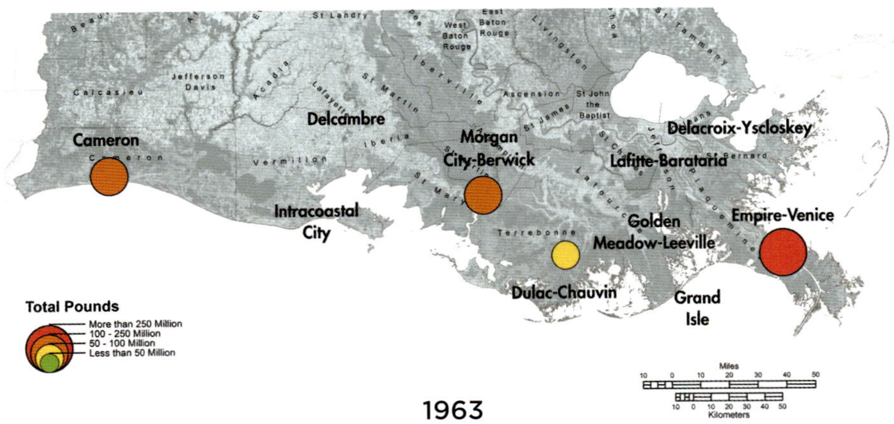

Fig. 33. Value of Fish and Shellfish Landed at Major Ports

6 maps

Data Sources: U.S. Department of the Interior, Bureau of Commercial Fisheries; NOAA, National Marine Fisheries Service

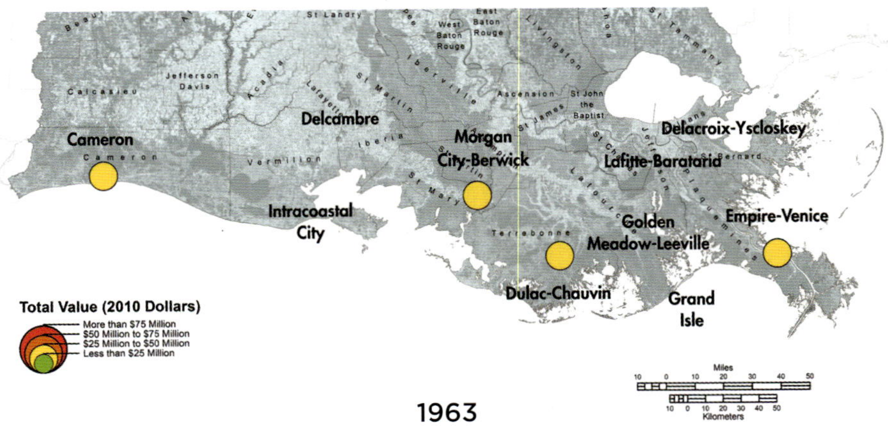

The secondary effects of many natural and technological hazards, outside of the immediate physical risk to life and property, are largely economic. The total numbers (fig. 32) and values (fig. 33) of fisheries landings, for example, are potentially impacted by hazard events. Played out against a background of natural shifts in aquatic species populations and global economic forces, several notable fluctuations in landings and values have occurred over the decades. In the 1970s, for example, the landings of menhaden, a high-yield, low-value industrial fishery, increased at ports in southwest and southeast Louisiana. These ports showed consistently high numbers of landings, although the values of these landings did not always correspond. One of the most dramatic impacts occurred following the Deepwater Horizon oil release of 2010. The number of landings remained high, even as lack of demand drove values down.

Fig. 32. Pounds of Fish and Shellfish Landed at Major Ports

Data Sources: U.S. Department of the Interior, Bureau of Commercial Fisheries; NOAA, National Marine Fisheries Service

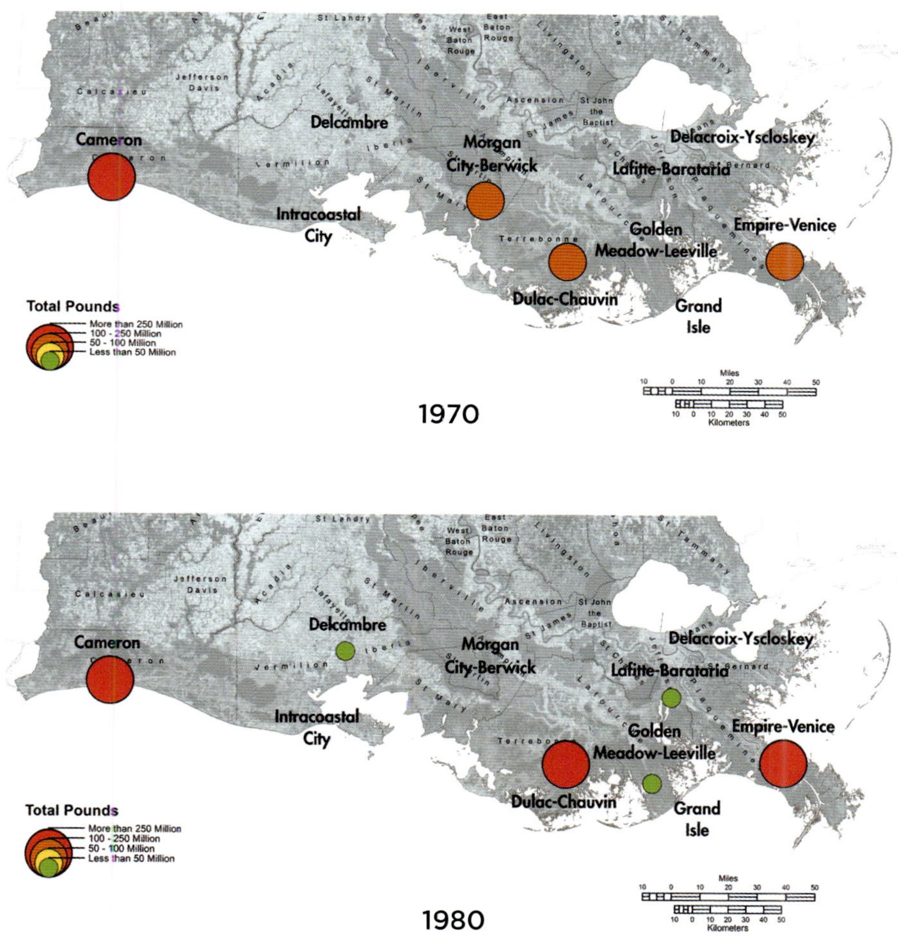

Fig. 33. Value of Fish and Shellfish Landed at Major Ports

Data Sources: U.S. Department of the Interior, Bureau of Commercial Fisheries; NOAA, National Marine Fisheries Service

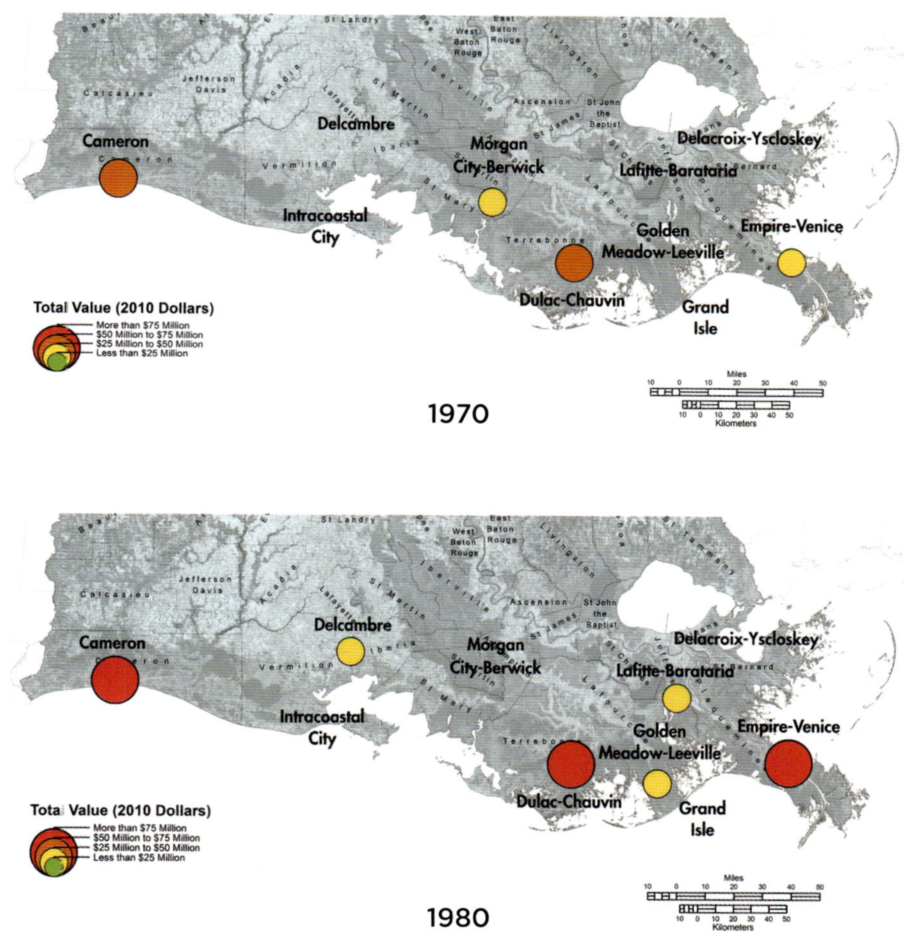

Fig. 32. Pounds of Fish and Shellfish Landed at Major Ports

Data Sources: U.S. Department of the Interior, Bureau of Commercial Fisheries; NOAA, National Marine Fisheries Service

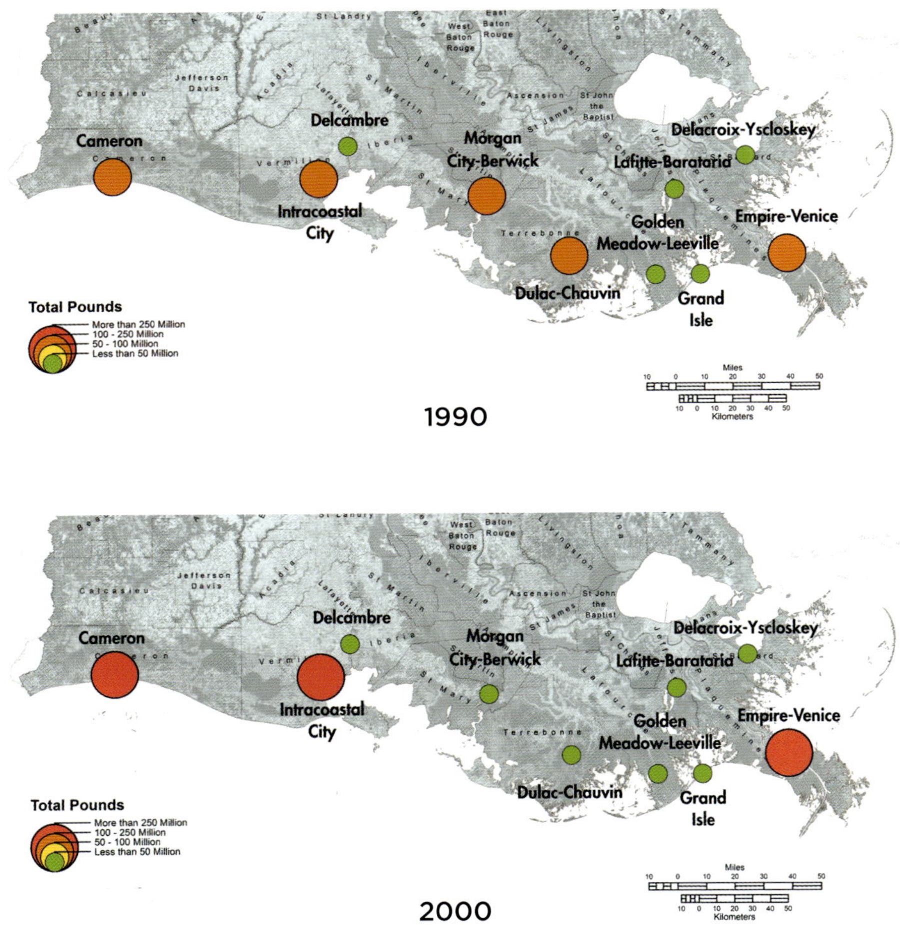

Fig. 33. Value of Fish and Shellfish Landed at Major Ports

Data Sources: U.S. Department of the Interior, Bureau of Commercial Fisheries; NOAA, National Marine Fisheries Service

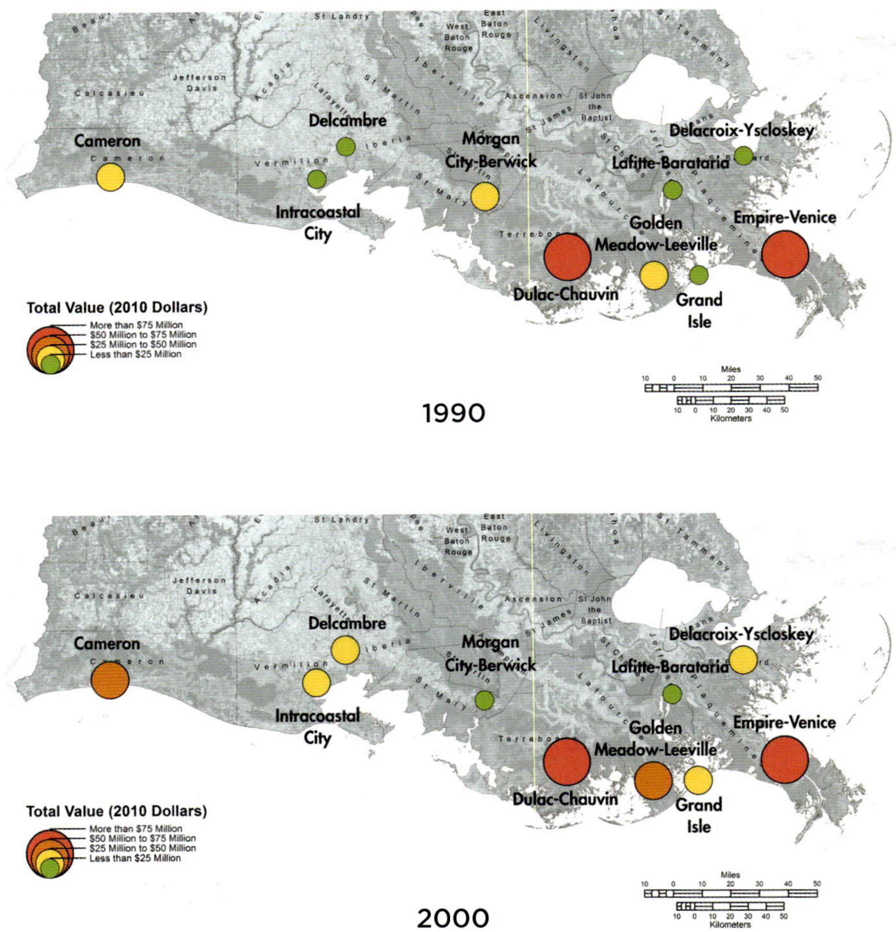

Fig. 32. Pounds of Fish and Shellfish Landed at Major Ports

Data Sources: U.S. Department of the Interior, Bureau of Commercial Fisheries; NOAA, National Marine Fisheries Service

Fig. 33. Value of Fish and Shellfish Landed at Major Ports

Data Sources: U.S. Department of the Interior, Bureau of Commercial Fisheries; NOAA, National Marine Fisheries Service

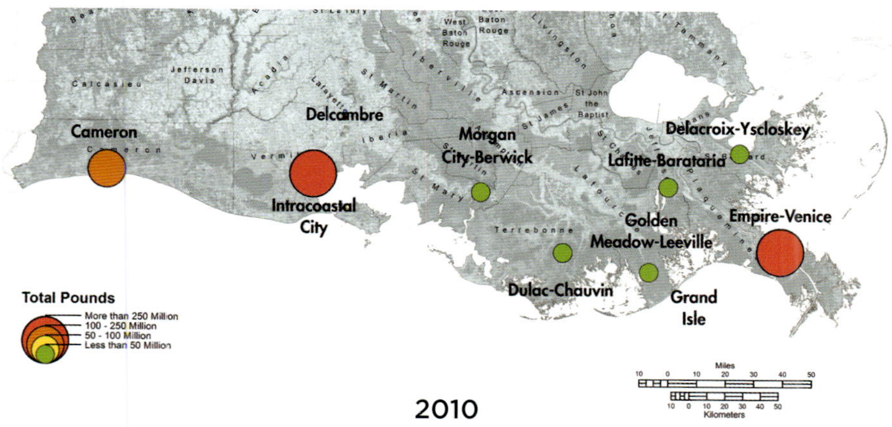

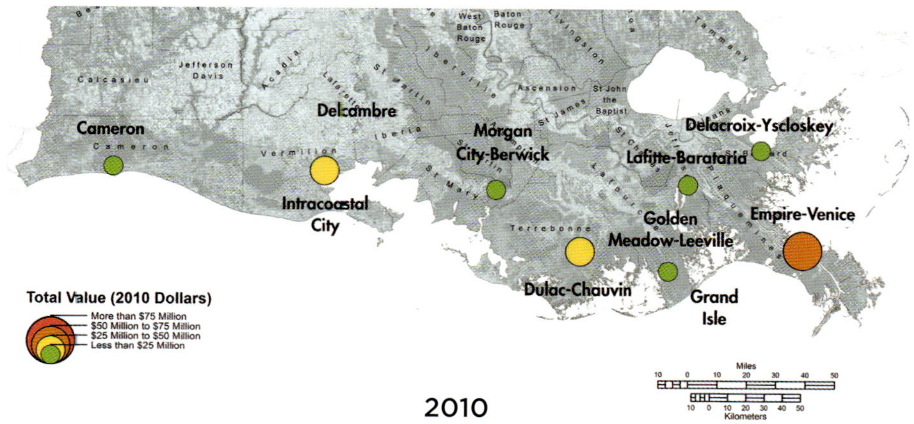

Fig. 34. Percentage of Agricultural Land in Coastal Shoreline and Coastal Watershed Parishes

8 maps

Data Source: U.S. Census of Agriculture

Fig. 35. Farms per Square Mile in Coastal Shoreline and Coastal Watershed Parishes

8 maps

Data Source: U.S. Census of Agriculture

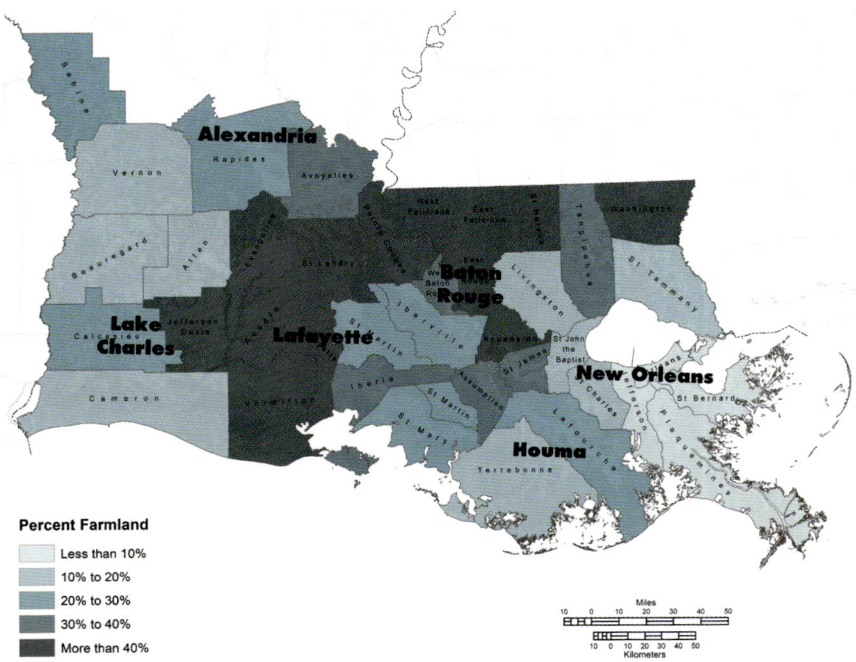

1945

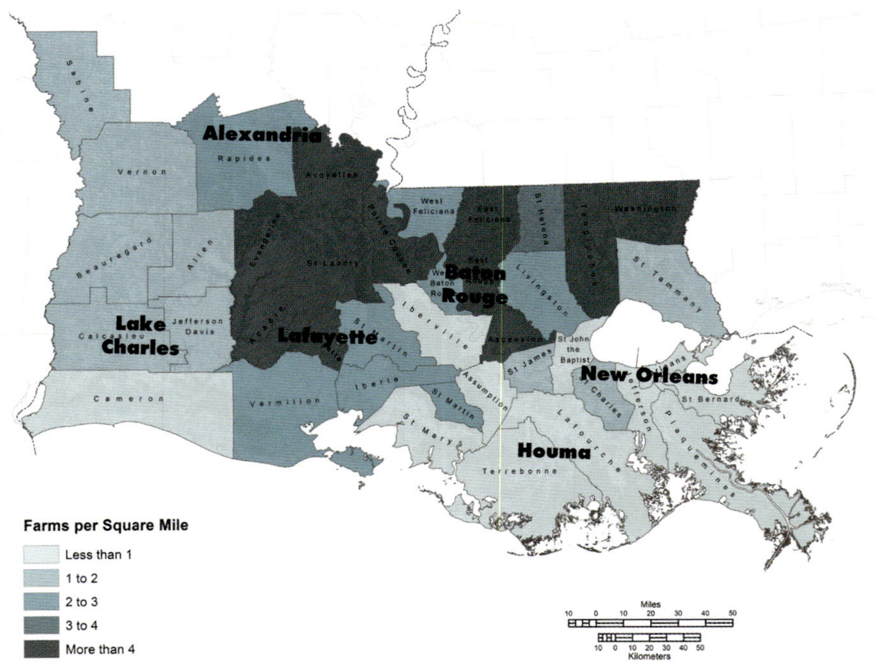

1945

As with fisheries, outside forces have tremendous impacts on agriculture. While the percentage of agricultural land in coastal parishes has remained relatively consistent over the last half of the twentieth century (fig. 34), the number of farms themselves has steadily declined (fig. 35). This suggests a shift toward large-scale industrial farming, with many smaller farms being incorporated into larger agglomerations. The long-term impacts of this shift on local communities is important to consider, as residents of small agricultural communities must transition into new economies.

Fig. 34. Percentage of Agricultural Land in Coastal Shoreline and Coastal Watershed Parishes

Data Source: U.S. Census of Agriculture

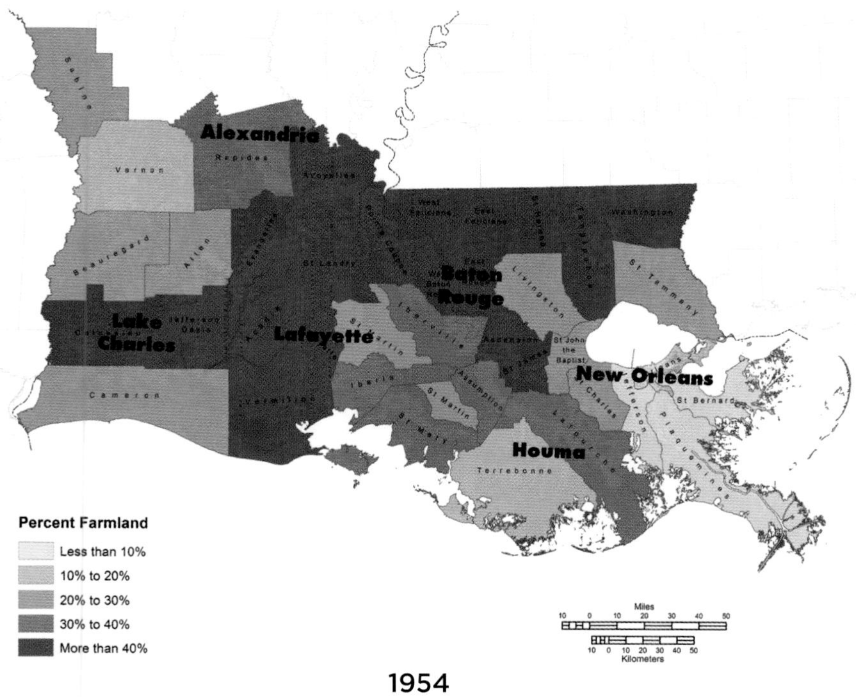

Fig. 35. Farms per Square Mile in Coastal Shoreline and Coastal Watershed Parishes

Data Source: U.S. Census of Agriculture

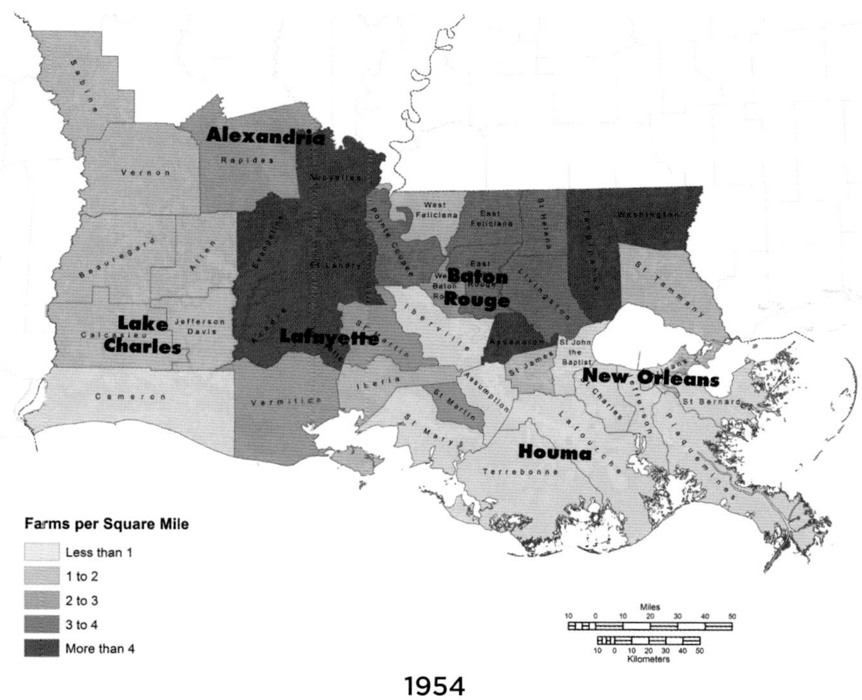

Fig. 34. Percentage of Agricultural Land in Coastal Shoreline and Coastal Watershed Parishes

Data Source: U.S. Census of Agriculture

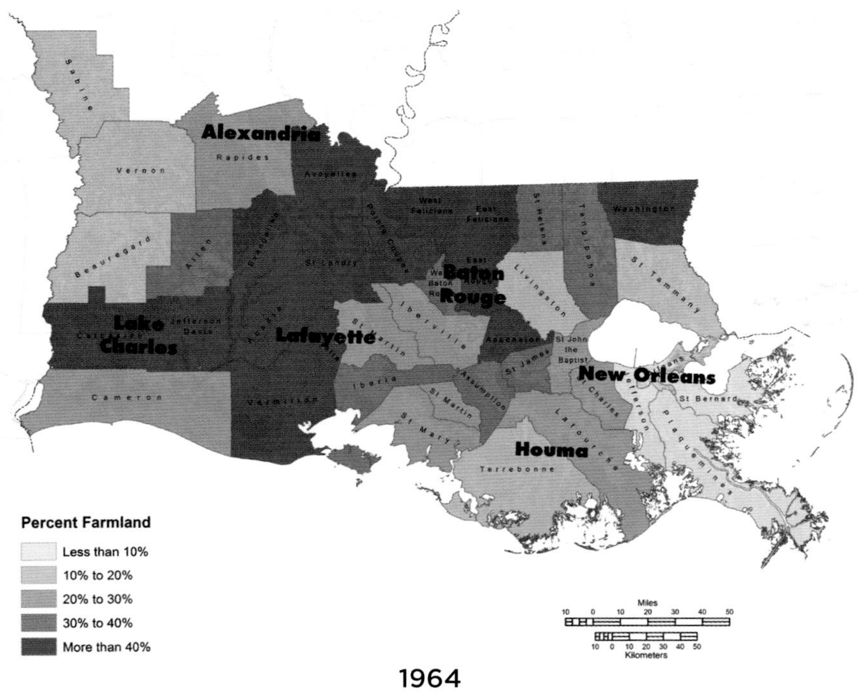

Fig. 35. Farms per Square Mile in Coastal Shoreline and Coastal Watershed Parishes

Data Source: U.S. Census of Agriculture

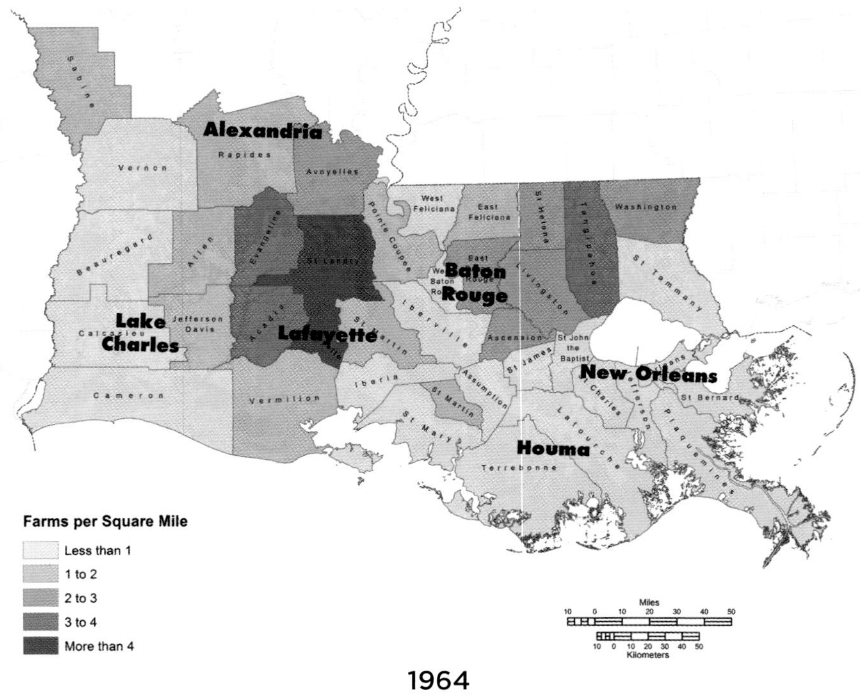

Fig. 34. Percentage of Agricultural Land in Coastal Shoreline and Coastal Watershed Parishes

Data Source: U.S. Census of Agriculture

Fig. 35. Farms per Square Mile in Coastal Shoreline and Coastal Watershed Parishes

Data Source: U.S. Census of Agriculture

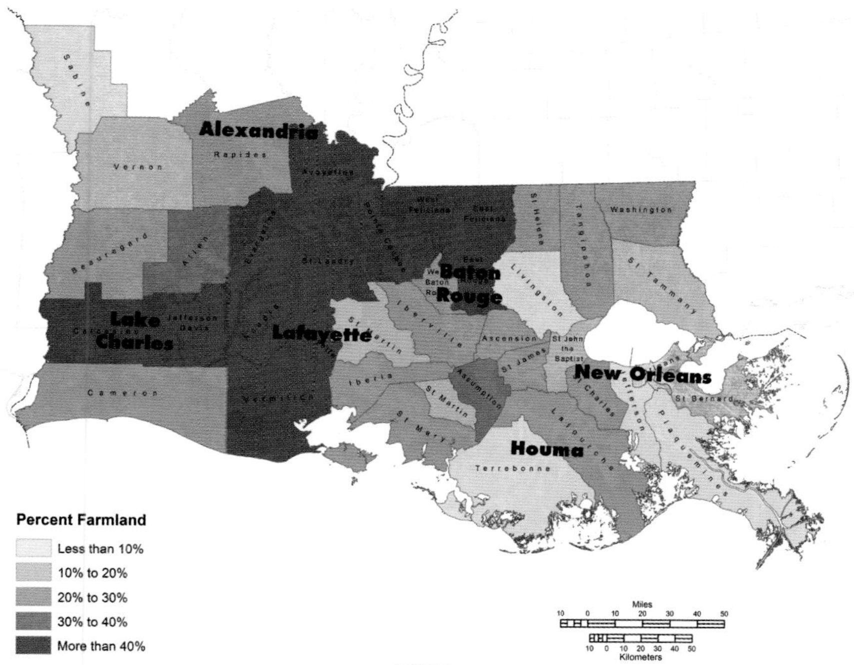

1974

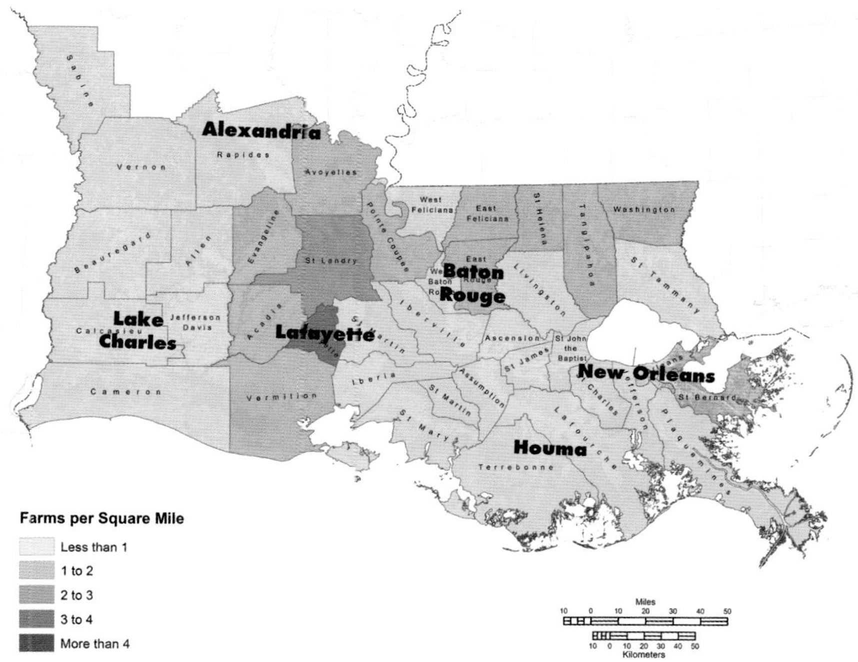

1974

Fig. 34. Percentage of Agricultural Land in Coastal Shoreline and Coastal Watershed Parishes

Data Source: U.S. Census of Agriculture

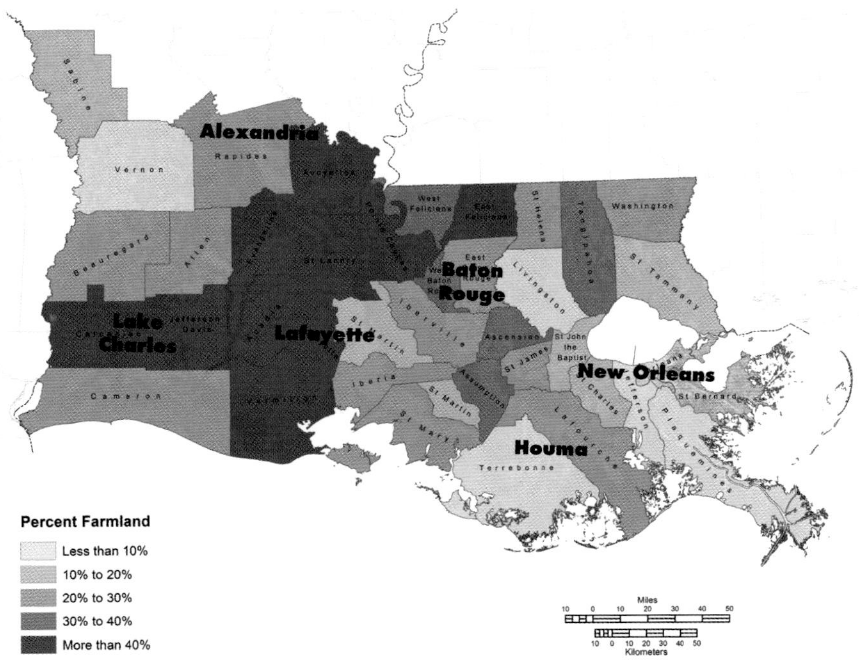

1982

Fig. 35. Farms per Square Mile in Coastal Shoreline and Coastal Watershed Parishes

Data Source: U.S. Census of Agriculture

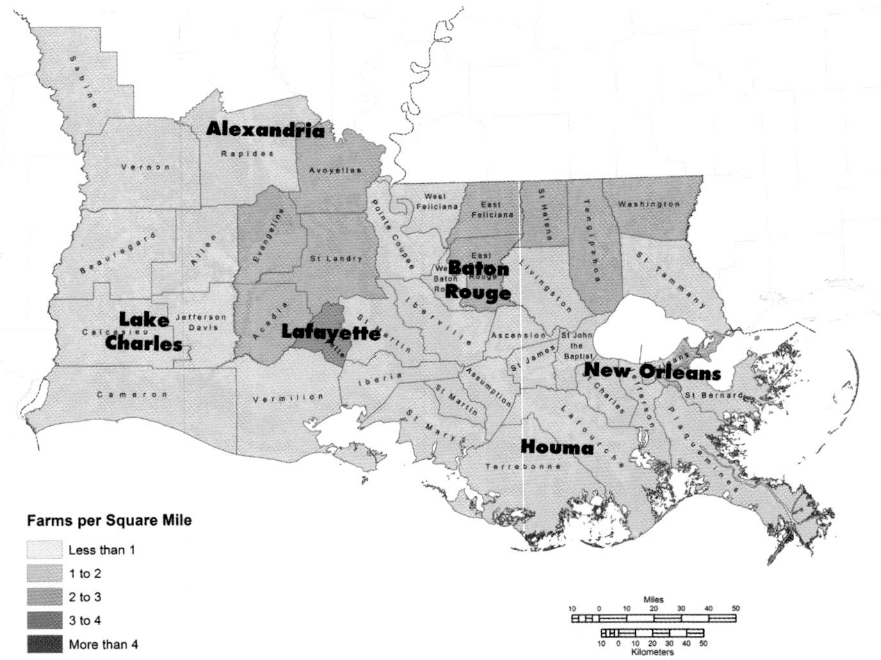

1982

Fig. 34. Percentage of Agricultural Land in Coastal Shoreline and Coastal Watershed Parishes

Data Source: U.S. Census of Agriculture

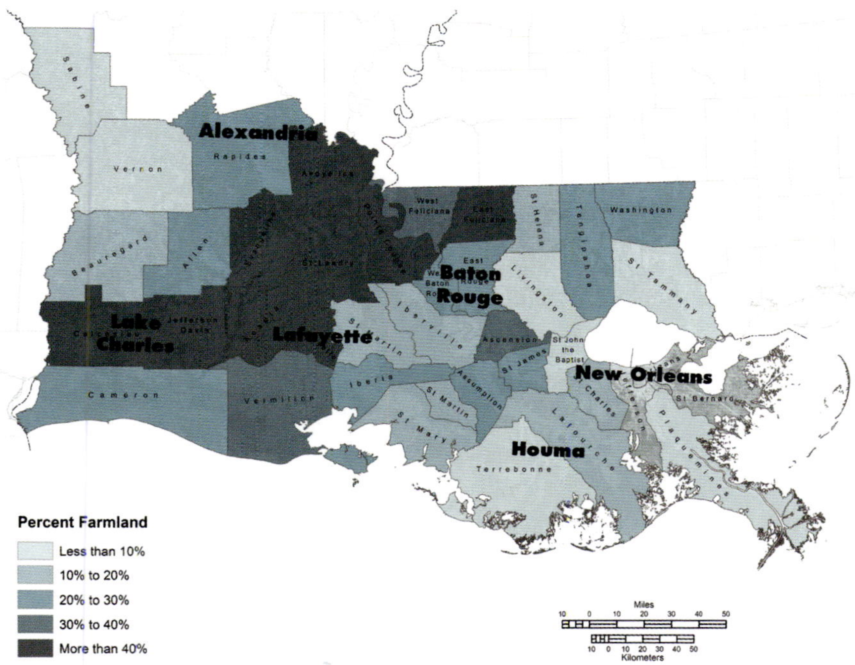

1992

Fig. 35. Farms per Square Mile in Coastal Shoreline and Coastal Watershed Parishes

Data Source: U.S. Census of Agriculture

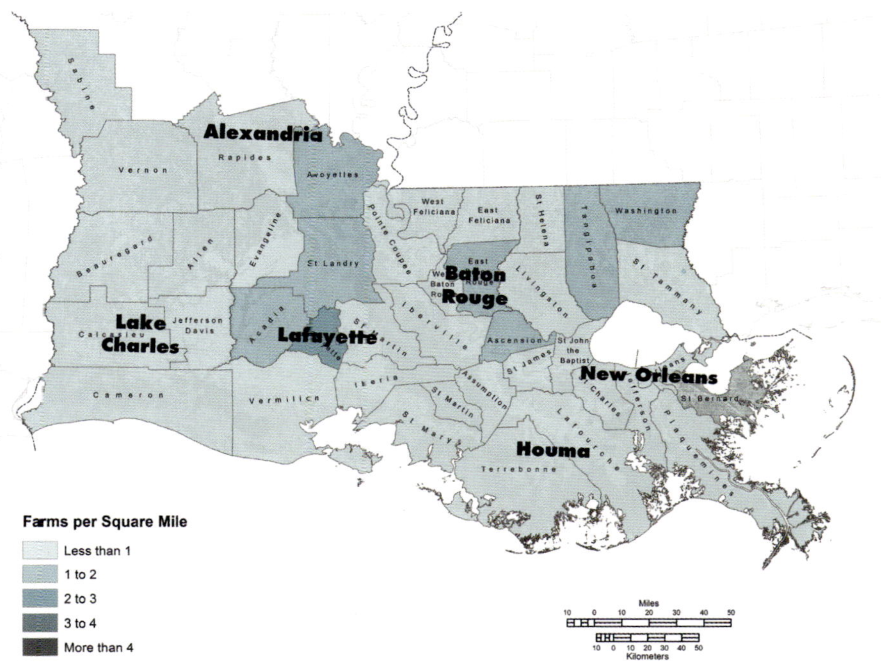

1992

Fig. 34. Percentage of Agricultural Land in Coastal Shoreline and Coastal Watershed Parishes

Data Source: U.S. Census of Agriculture

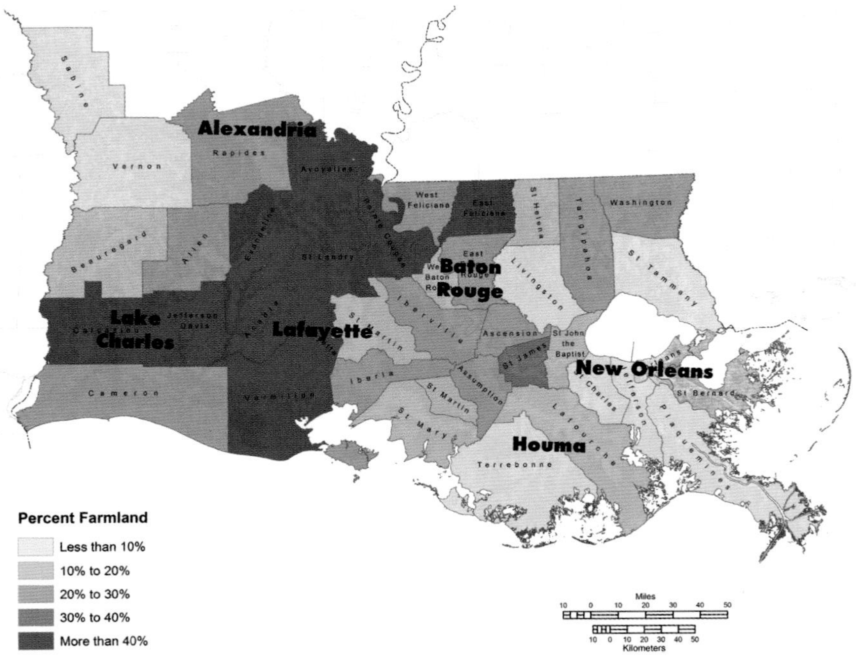

2002

Fig. 35. Farms per Square Mile in Coastal Shoreline and Coastal Watershed Parishes

Data Source: U.S. Census of Agriculture

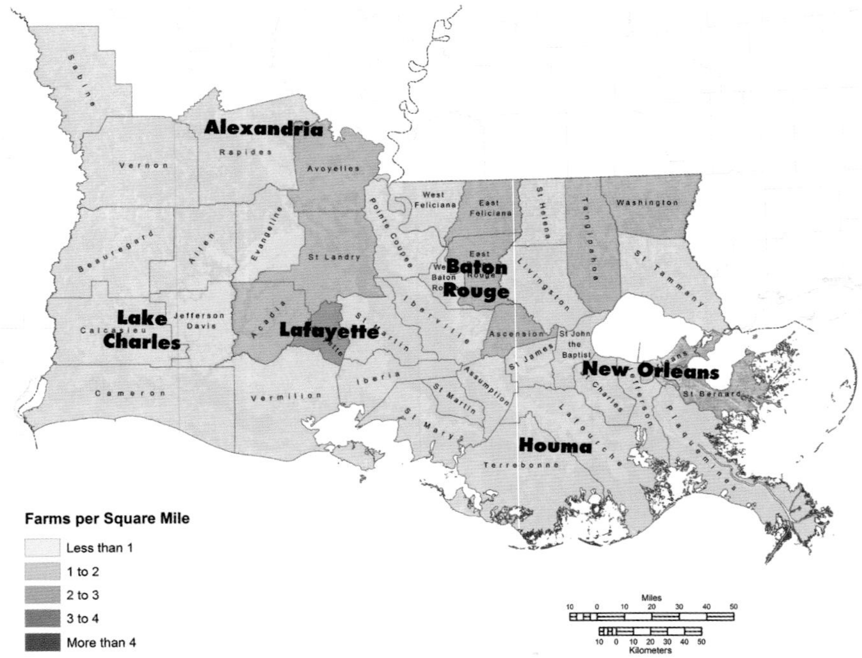

2002

Fig. 34. Percentage of Agricultural Land in Coastal Shoreline and Coastal Watershed Parishes

Data Source: U.S. Census of Agriculture

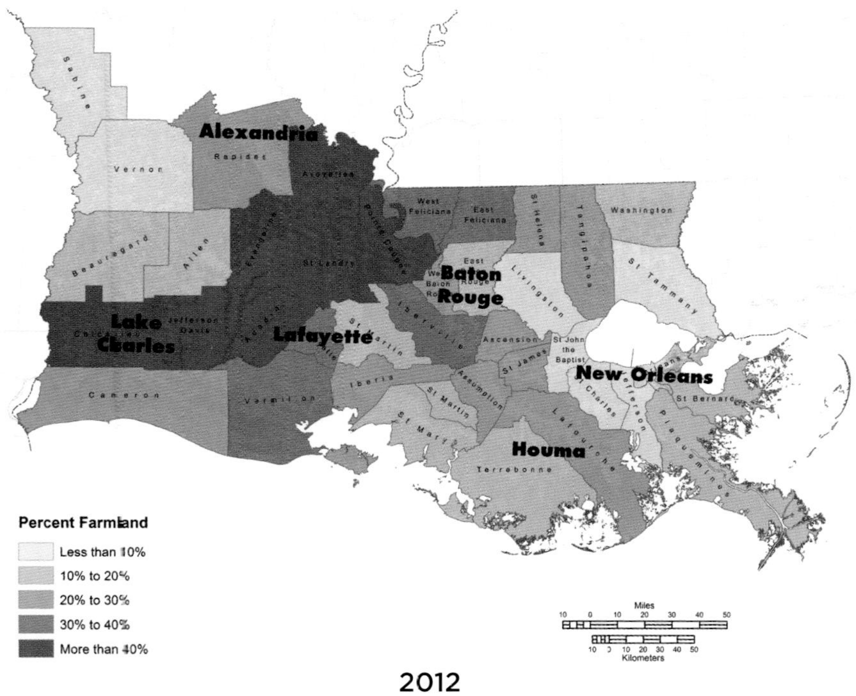

Fig. 35. Farms per Square Mile in Coastal Shoreline and Coastal Watershed Parishes

Data Source: U.S. Census of Agriculture

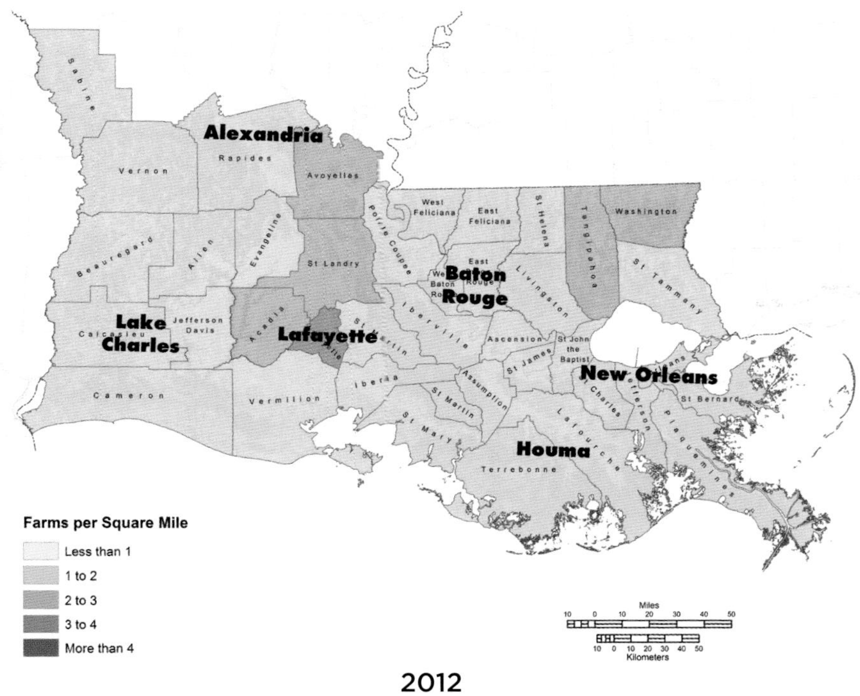

RISKS ENHANCED THROUGH ENVIRONMENTAL MODIFICATION

Hazards that combine natural and human-induced factors present another aspect of risk in coastal Louisiana—risk enhanced through environmental modification. Examples of this type of hazard include the landscape-level changes wrought by the construction of navigation channels, pipeline canals, and logging canals.

Fig. 36. The Influence of Waterways on Storm Surge during Hurricane Katrina
Data Source: LSU Coastal Emergency Risks Assessment

Navigation channels, like most other open waterways, have the potential to funnel storm surge far inland. The Mississippi River-Gulf Outlet (MR-GO) is an artificial navigation channel designed to provide oceangoing vessels with a more direct route to the Gulf of Mexico than that provided by the Mississippi River. The effect of the Gulf Intracoastal Waterway (GIWW) and MR-GO on storm surge during, among other events, Hurricane Katrina (fig. 36) was profound. As Katrina's storm surge moved toward southeast Louisiana, across the Breton, Chandeleur, and Mississippi sounds and into Lake Borgne, surge pressure intensified on the Inner Harbor Navigation Canal (the Industrial Canal) on the eastern edge of New Orleans. Ultimately, breaches and overtopping of levees occurred along the GIWW, the MR-GO, and the Industrial Canal, resulting in flooding across the Lower Ninth Ward of New Orleans, New Orleans East, and St. Bernard Parish.

Fig. 37. Hydrological and Ecological Changes in Selected Locations Proximate to the MR-GO

3 maps

Image Sources: USGS 1998 DOQQ Imagery; USGS 2012 DOQQ Imagery; USGS Historical Topographic Quadrangle Maps

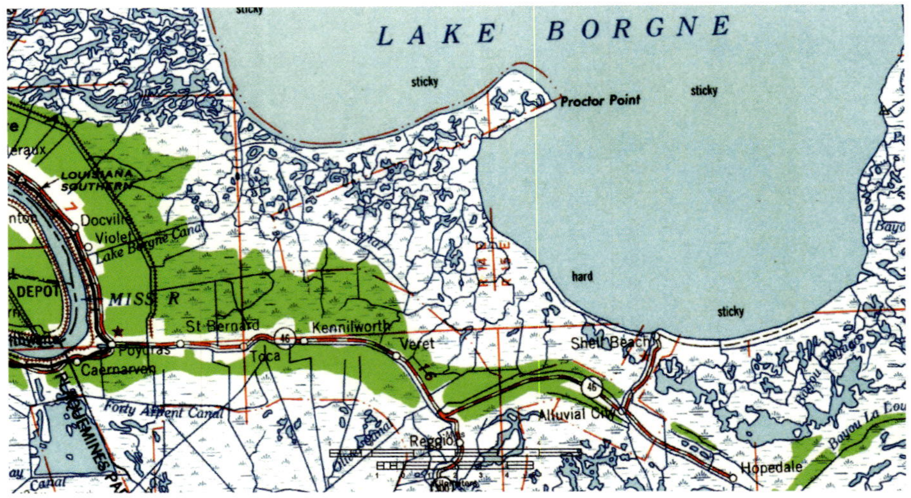

1961

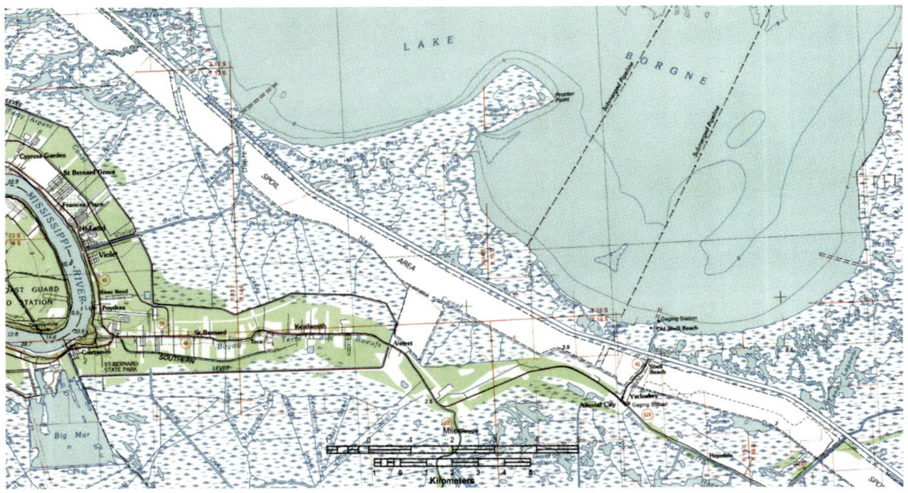

1983

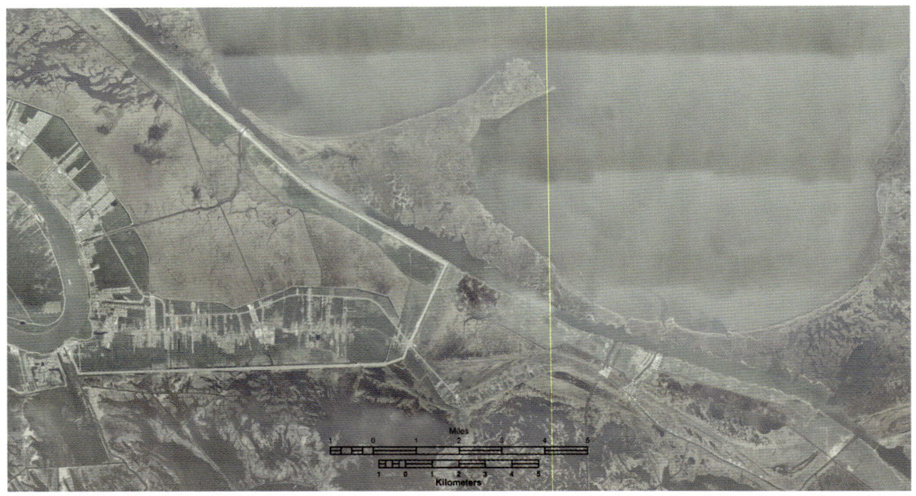

2012

Other impacts of these channels are hydrological in nature, effectively fragmenting the landscape and cutting off surrounding wetlands from their water sources, ultimately degrading the wetlands (fig. 37). For example, over time, most of the intermediate marsh and cypress swamps proximate to the MR-GO have converted to brackish marsh while previously brackish marsh has converted to saline marsh. These changes have had profound effects on commercial fisheries as well as waterfowl and mammal habitat. These impacts are insidious, often occurring over years, if not decades.

Fig. 38. The Density of Pipeline Crossings in Coastal Louisiana

Data Source: Louisiana Oil Spill Coordinator's Office

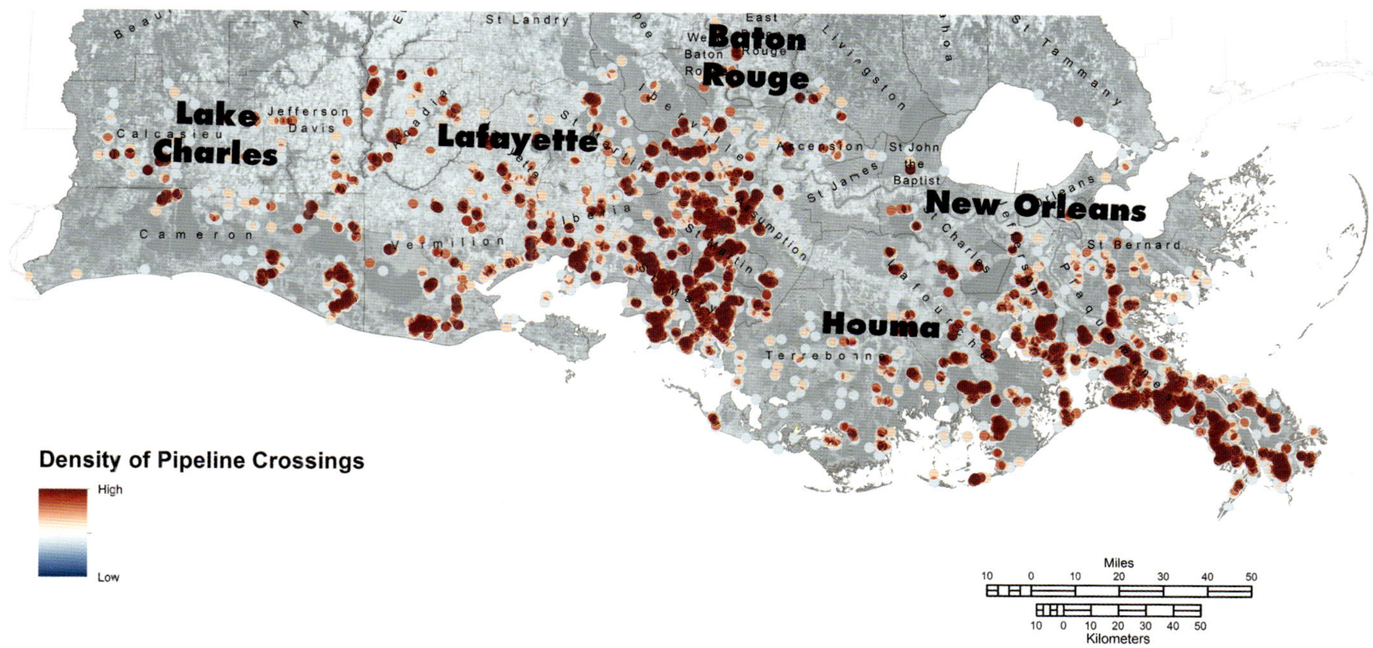

Unlike the direct, immediate impacts of the funneling of storm water through open waterways, pipeline and logging canals also increase community vulnerability and cause long-term degradation of the landscape. The locations where oil and gas pipelines cross roads, rivers, and bayous (fig. 38) are some of the most susceptible locations for both pipe corrosion and third-party damage, both of which represent an increased risk of oil spills.

Fig. 39. Fracturing the Landscape: The Historical Impacts of Pipelines

6 maps

U.S. Geological Survey quadrangle maps and aerial photography show the development of the Lafitte Oil Field in Jefferson Parish as well as the transformation of the landscape wrought by the extensive development of canals in the area. By the twenty-first century, much of what was once coastal wetland had converted to open water.

Image Sources: USGS 1998 DOQQ Imagery; USGS 2012 DOQQ Imagery; USGS Historical Topographic Quadrangle Maps

1948

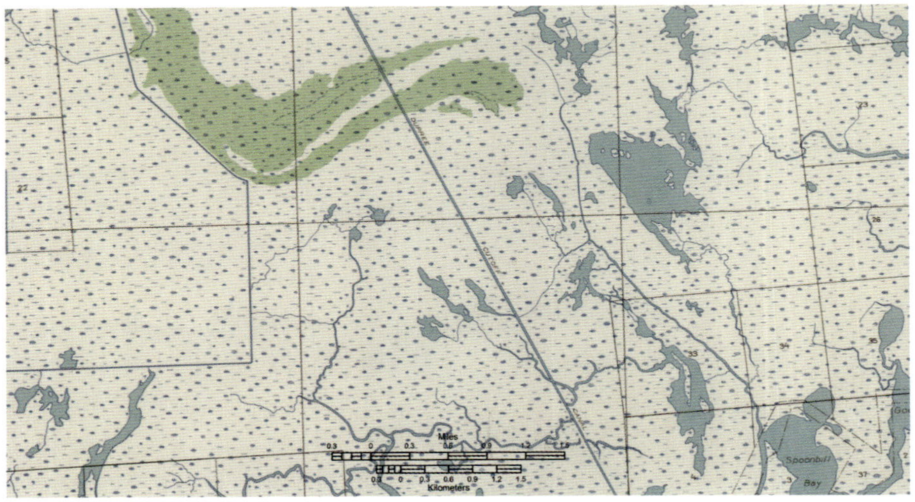

1932

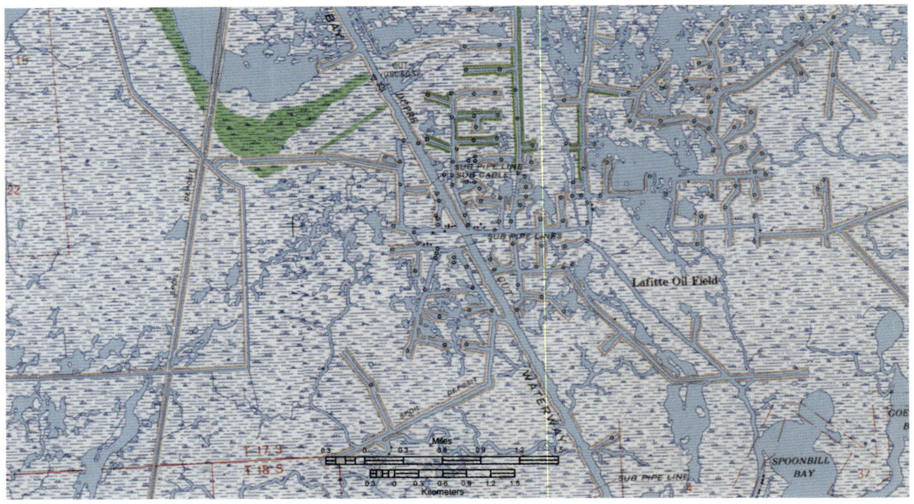

1962

Oil and gas canals were used historically in coastal Louisiana to transport exploration and production equipment as well as to house pipelines. These canals, like navigation channels, were generally dredged, and the sediment dug up by this process was deposited as spoil banks alongside the canal, often filling in the surrounding wetlands. While this immediate loss of wetlands is significant, the greatest damage occurs over time (fig. 39). The opening of the wetlands to enhanced wave action, both natural and induced by the wake from boat traffic, has hastened their loss, severely degrading the coastal landscape.

Fig. 39. Fracturing the Landscape: The Historical Impacts of Pipelines

Image Sources: USGS 1998 DOQQ Imagery; USGS 2012 DOQQ Imagery; USGS Historical Topographic Quadrangle Maps

1998

1973

2012

Fig. 40. Permanent Scars on the Landscape: Historical Cypress Logging Canals

3 maps

Image Sources: USGS 2012 DOQQ Imagery; USGS Earth Resources Observations and Science Center

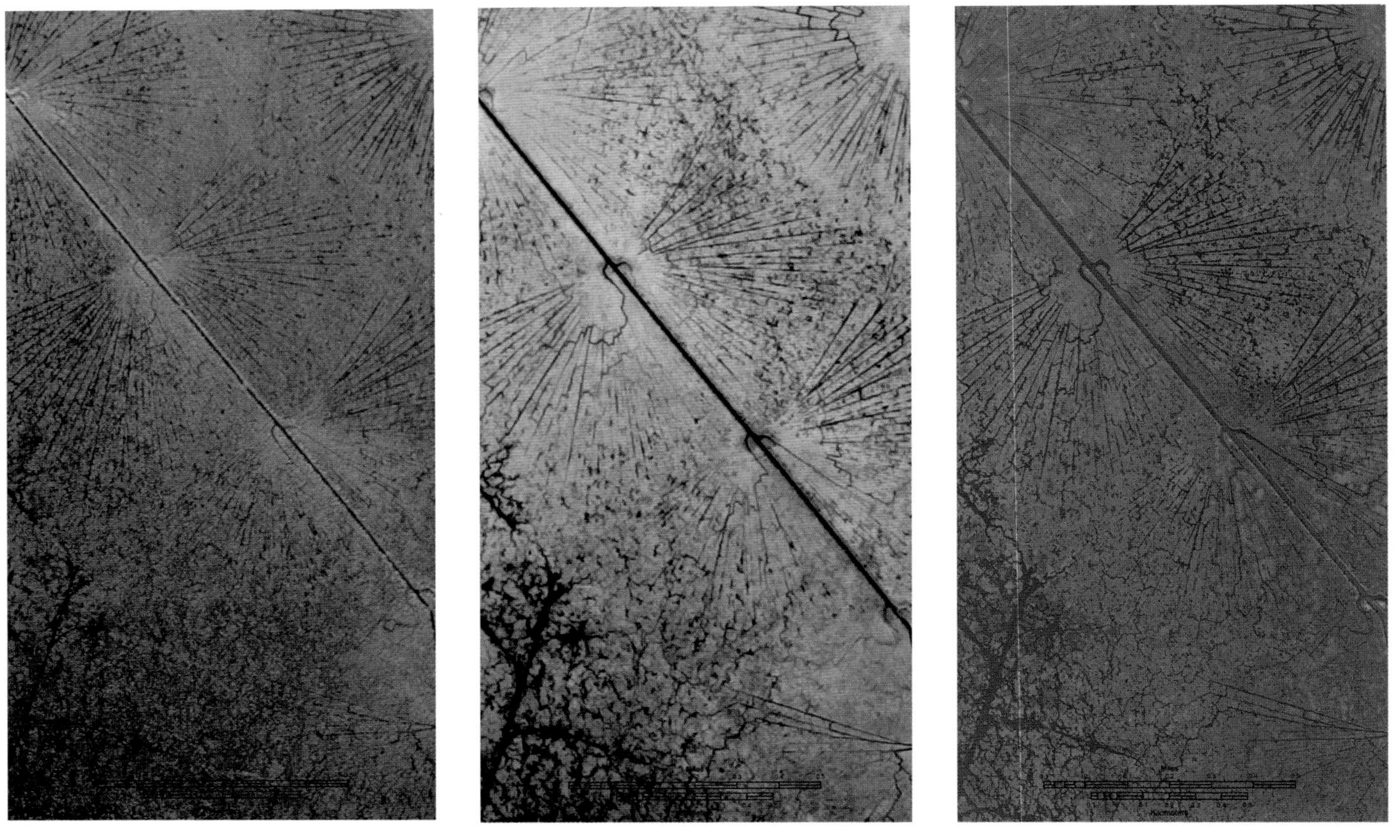

1965 1978 2012

Logging canals have had similar degrading impacts on the environment. Historical methods of cypress logging involved cutting a series of radial pulling runs through which cypress logs could be dragged to a central location for transport to the sawmills. These pullboat runs were generally cleared of trees, with the stumps often blown out with dynamite. The process of dragging cypress logs through these runs dug deep ditches in the landscape that made regrowth difficult. These canals persist in the landscape as evidence of the ecological changes caused by development in the coastal zone (fig. 40).

CHAPTER 3

REBUILDING THE FOUNDATION

Historical Adaptations to Internal and External Shocks

LOUISIANA'S COASTAL ZONE is a complex system driven by a range of shifting ecological and geophysical conditions. Centuries of human-induced modifications to the landscape have effectively tied social relations to physical processes in intricate ways, creating dependencies between society and ecology that are difficult to disentangle analytically or politically (Dalbom et al. 2014). The coupled human-natural system that is coastal Louisiana has had to constantly adapt to life on the soft edge of the continent where environmental risks are ever present. The degree to which societies are able to effectively cope with and adapt to changing conditions is a key factor in determining how resilient that society is. This is true at any scale within society, from the individual to the community to the region. Social vulnerability is present in all populations, to some degree. It develops and changes over time as communities adapt in response to any number of triggering events.

There is a common belief that adaptation is the result of deliberate policy decisions designed to return to, maintain, or achieve some desired state, presumably one of resilience (Noble et al. 2014). Indeed, institutional, policy-driven adaptation is a primary process of risk reduction in coastal areas, particularly as it relates to transforming the landscape into a setting that is capable of sustaining those communities that depend on it. Similarly, policy adaptation can be used to reduce exposure to risk by limiting the amount of development in hazardous areas. These types of deliberate adaptation measures are generally implemented on short time scales, relatively speaking. For this reason, they can be considered to be discrete, short-term episodic events, with impacts that can be measured in a single decade. Many of these adaptations have been technological or engineered. Intricate networks of physical infrastructure attempt to stabilize coastal and deltaic processes and enable human settlement, mobility, and economic production in this inherently unstable landscape. Other policy-related adaptive strategies are more behavioral or institutional in nature, attempting to modify the practices of individuals, groups, or institutions (Smithers and Smit 1997). Such strategies do not attempt, in general, to modify the biophysical environment itself, but rather to alter the relationship that the human community has with the natural.

Whatever planned adaptive strategy society undertakes to adjust to shifting environmental conditions does not occur in isolation from the influence of other intrinsic economic, social, and institutional forces. While adaptation is commonly thought of as a public policy response to environmental change and increasing risk levels, potentially affected communities or regions will often adapt spontaneously to changes in environmental conditions, often quite apart from policy (Smithers and Smit 1997). This type of inherent adaptation often results in demographic or economic modifications to the structure of the community or region, such as broad-scale shifts in regional economic activity or changing patterns of population settlement. Such internal changes are generally slow moving and may take several decades or even generations to effectively identify and understand the impacts on the community and region.

Just as economic, social, and institutional forces may impede or mediate changes in human systems, there are also examples of these nonenvironmental factors resulting in the development of maladaptative practices—those adaptations that ultimately increase social vulnerability to environmental change. Such maladaptations are driven largely by nonclimatic forces and may serve short-term human goals, but often with attendant costs on individuals, communities, and society (Smithers and Smit 1997). Maladaptations may be guided by policy, such as providing insurance subsidies to encourage development of flood zones and other high-risk areas, or by economics, such as the construction of slab-on-grade housing in floodplains or the planting of moisture-sensitive crops in drought-prone areas.

Despite the best efforts by all levels of society to adapt, a combination of ill-suited development, changing environmental conditions, and land loss across coastal Louisiana has resulted in increased environmental, economic, and social vulnerability. These vulnerabilities have been compounded by an increasing number of natural and technological hazard events, including hurricanes, river floods, and oil spills (Peyronnin et al. 2013). In the face of this increasing vulnerability and risk, the most fundamental adaptation decision that people must make is whether to remain in their homes and rely on protective measures to hold back the sea or to relocate out of harm's way. Both choices have important costs and uncertainties for the individual, the community, and society at large.

Population movement and relocation can reduce or eliminate some of the costs and environmental impacts of constructing massive shoreline protection systems, but often at the expense of the loss of homes and possibly entire communities and local culture. Prior to the twentieth century, high mobility was a successful adaptation strategy in many unstable environments, and population movement was traditionally a first response to various environmental pressures (Marino 2012). Through time, however, decreases in mobility coinciding with government strategies encouraging development within flood zones have resulted in situations where much of the population finds it difficult, if not impossible, to relocate. Migration and relocation are highly influenced by the degree of social vulnerability of the impacted community. The percentage of the population living in poverty, for example, is a direct measure of the community's ability to both evacuate and relocate when faced with a natural hazard event (Dalbom et al. 2014; Levine et al. 2007). Displacement and relocation in Louisiana's coastal plain have largely been driven by shifting patterns of flood disturbance and exposure in the urban periphery and coastal zone, those developing and lightly developed areas located outside of the structural flood protection systems. Until the decision to protect or relocate coastal populations is made, preparing for long-term sea-level rise in coastal Louisiana may be impossible (Titus et al. 2009). This decision will ultimately be made at the individual level, in spite of policy-driven efforts, and will be guided in large part by the lived experiences of the community, the effectiveness of previous adaptation efforts, and the level of inherent resilience possessed by that community.

HISTORICAL ADAPTATION TO RISK

Despite all the risks and hazards faced by residents of coastal Louisiana through the years, communities have persisted. There are two general means of adapting to life in the coastal zone. People adapt to risk and changing environmental conditions, or they modify the risk itself.

Fig. 41. Head of Cattle per Square Mile

4 maps

Data Sources: U.S. Census Bureau; USDA, National Agricultural Statistics Service

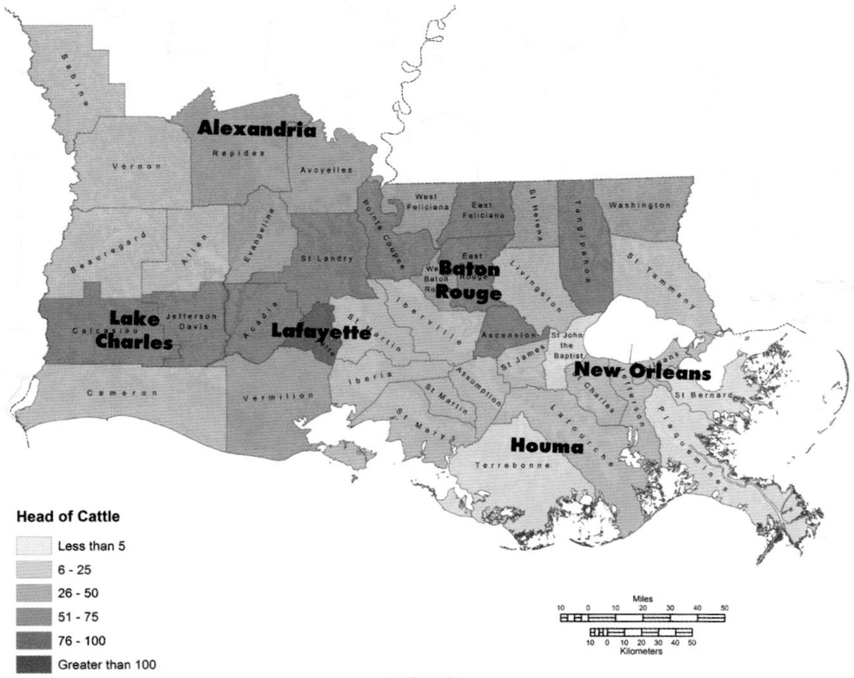

1950

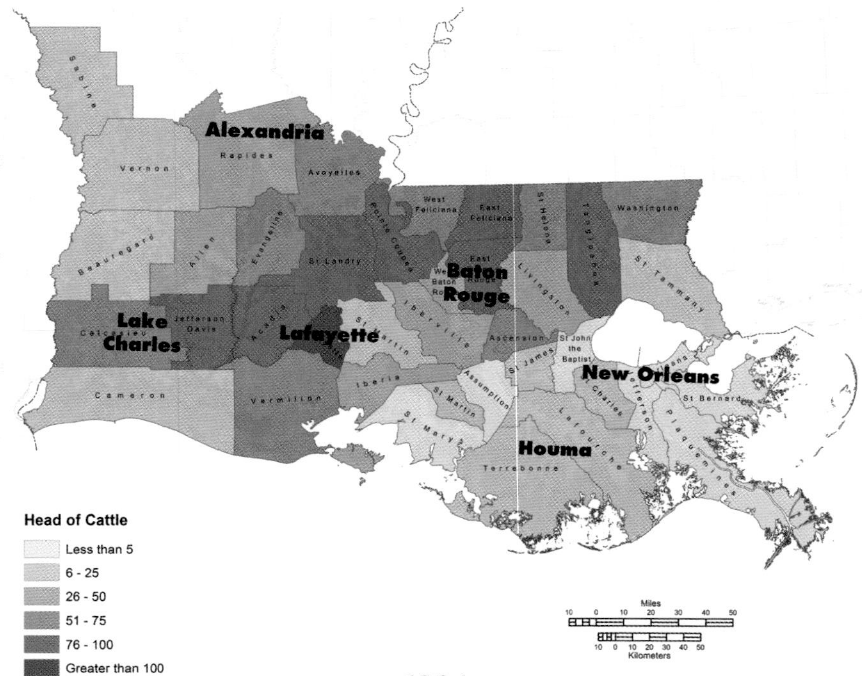

1964

Communities persist in coastal Louisiana in large part because this area is a working coast intricately tied to the environmental goods and amenities produced in the region. This linkage between communities and various ecosystem goods and services binds community to the environment. Communities have had to alter this relationship in varying ways. For example, cattle ranching in coastal Louisiana (fig. 41) historically involved a seasonal migration of cattle from ranches north of the coastal zone down to the coastal marshes and cheniers and back again. This migration allowed cattle ranchers to take advantage of the plentiful forage grasses in the marshes during the winter. Over time, however, the development of agriculture transformed the uplands of the chenier plain into a major rice-growing region. Today, cattle farmers in southwest Louisiana often grow rice and rotate their cattle through the rice fields, allowing the cattle to clean up the rice stubble and fertilize the fields (Adams 2013).

Fig. 41. Head of Cattle per Square Mile

Data Sources: U.S. Census Bureau; USDA, National Agricultural Statistics Service

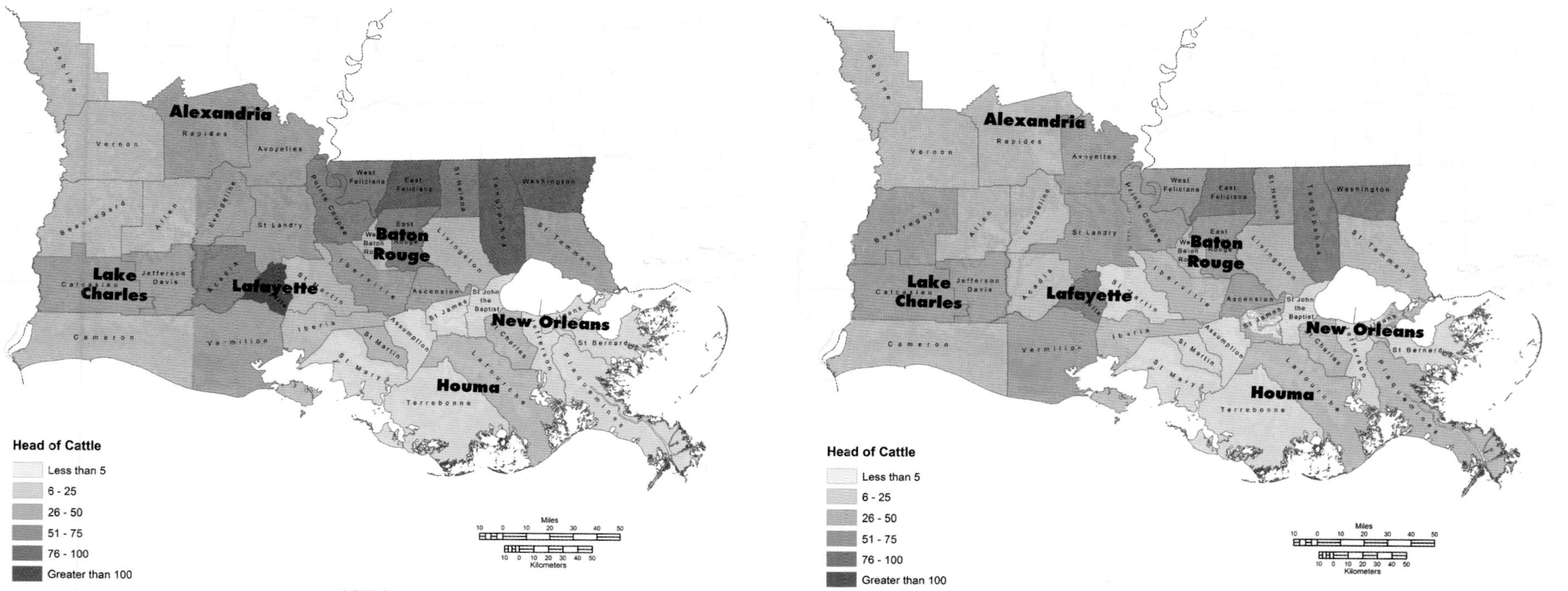

1974 1997

Fig. 42. Rice Production as a Proportion of Total Acres of Farmland

4 maps

Data Sources: U.S. Census Bureau; USDA, National Agricultural Statistics Service

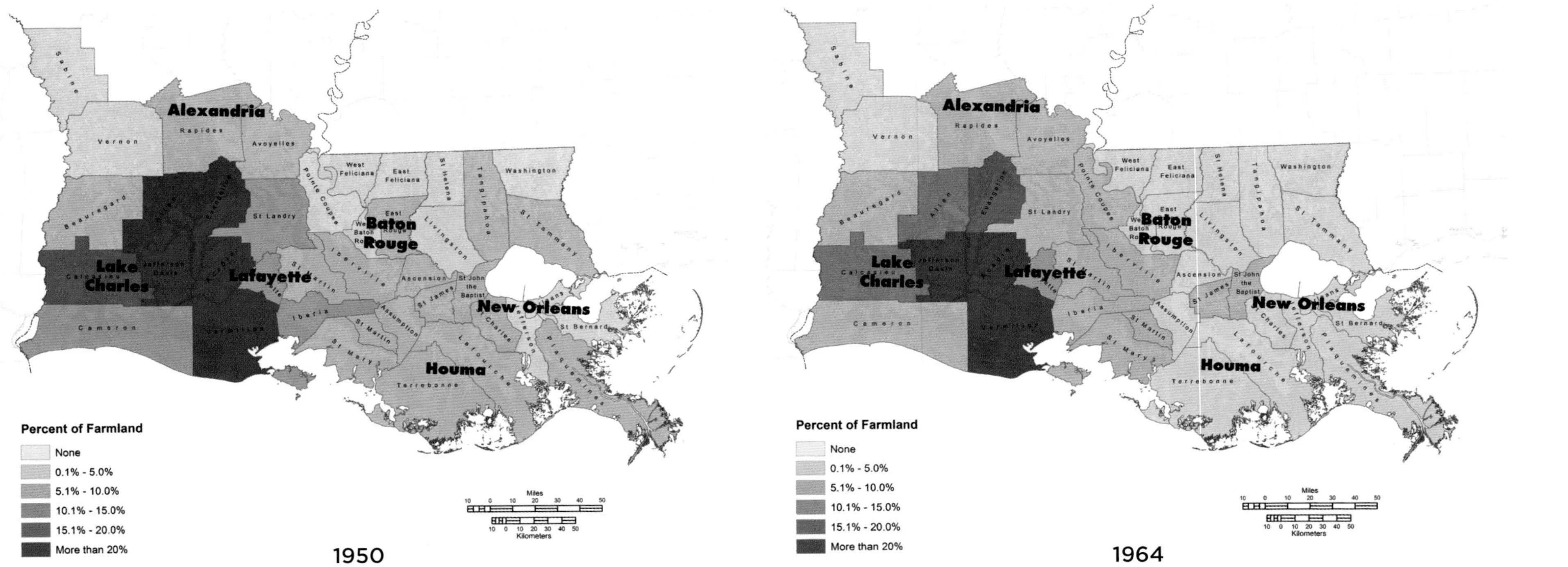

Oftentimes, adaptation to changing conditions has meant altering the crops grown in a region. Ideal land for rice production must be uniformly level and graded and also have access to large amounts of freshwater to flood the fields. Prior to the twentieth century, rice was grown in the parishes along the Mississippi River between New Orleans and Baton Rouge. Mechanization and prohibition of sluices in river levees were adaptations that shifted rice production to the coastal prairies. Since 1900, the rice production regions of the state have remained fairly stable (fig. 42).

Fig. 42. Rice Production as a Proportion of Total Acres of Farmland

Data Sources: U.S. Census Bureau; USDA, National Agricultural Statistics Service

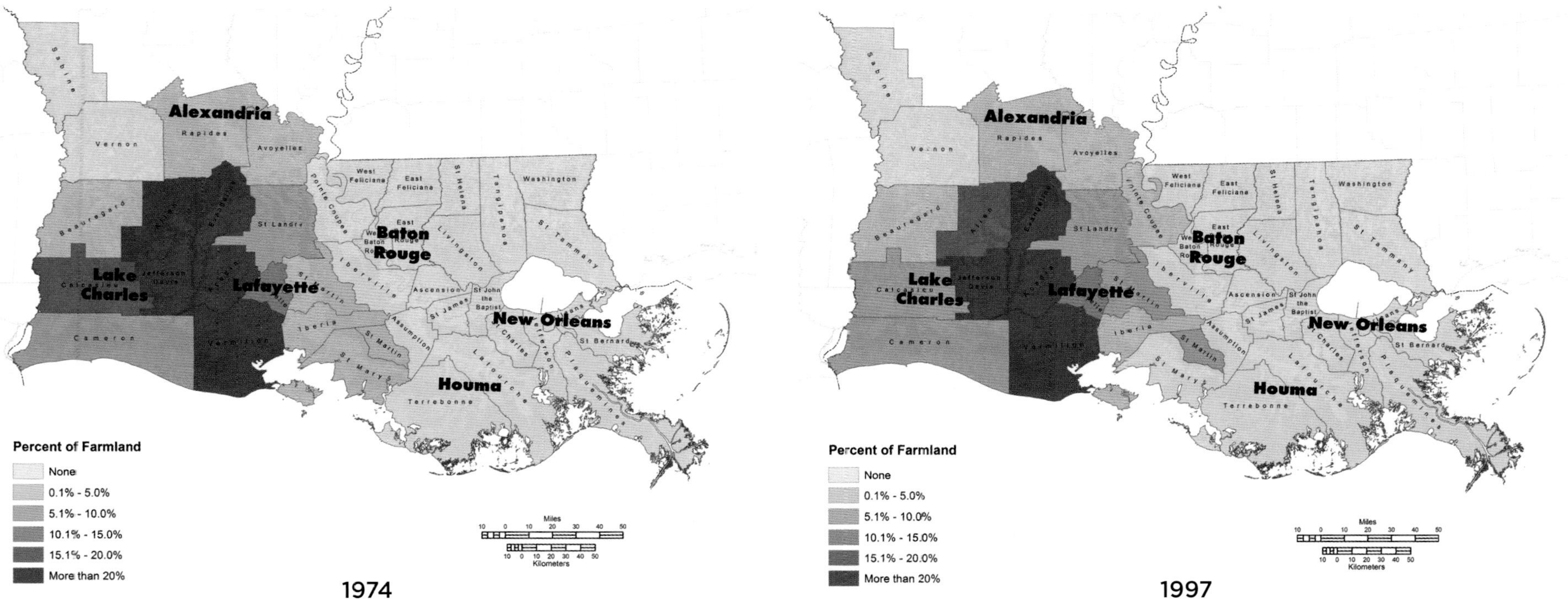

Fig. 43. Sugar Production as a Proportion of Total Acres of Farmland

4 maps

Data Sources: U.S. Census Bureau; USDA, National Agricultural Statistics Service

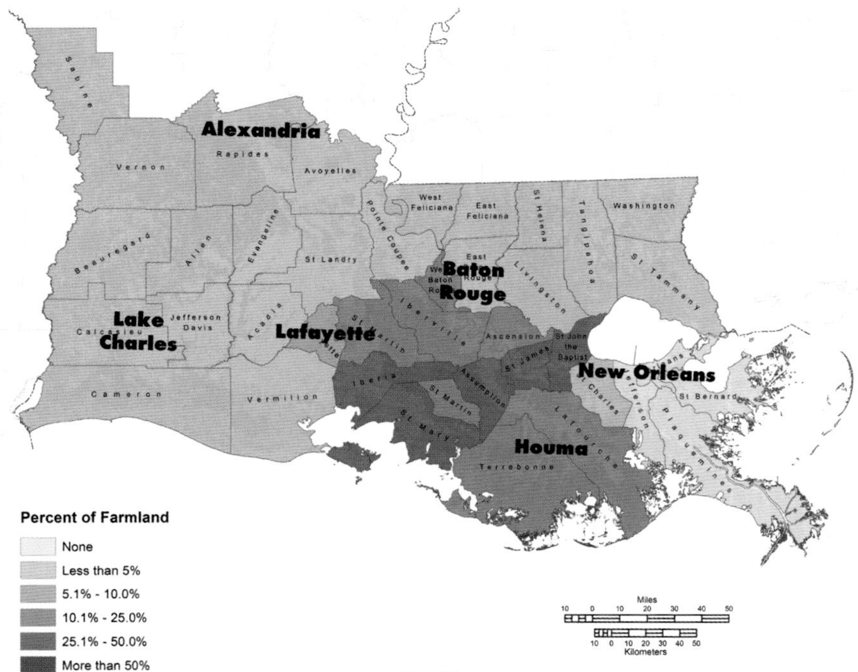

1950

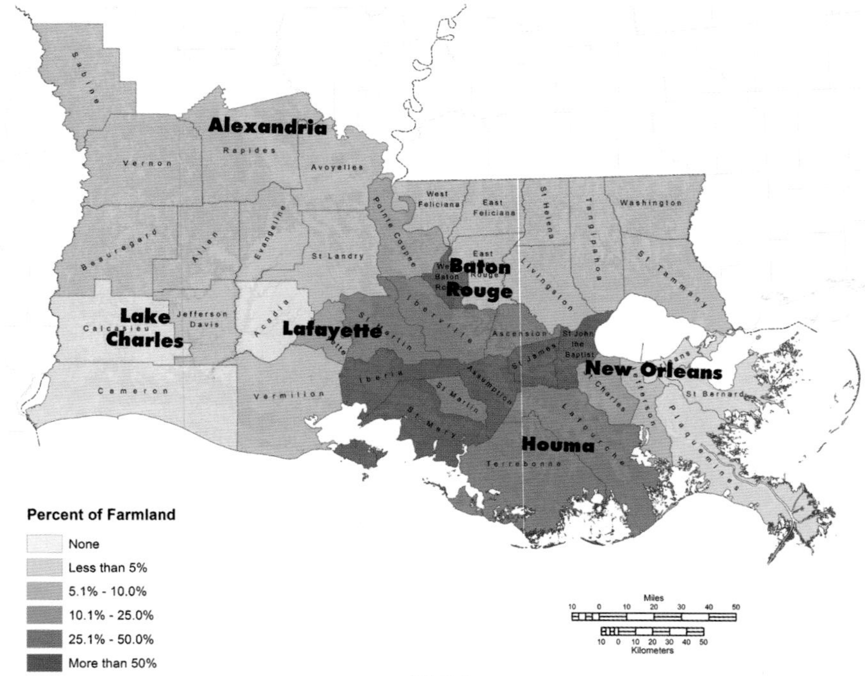

1964

Sugarcane is Louisiana's highest-value crop. Due in part to government subsidies, this crop has become more important in terms of relative acres grown per parish (fig. 43). The state's temperate climate and incidence of frost are major limitations on sugar production, resulting in a short growing season relative to tropical areas outside the United States. However, sugarcane breeding efforts have allowed Louisiana's sugar yields to continue to climb throughout the twentieth century. Historically, sugar has been cultivated along the natural levees of the Mississippi River, Bayou Lafourche, and the Red River. While this region remains the core of sugarcane production in south Louisiana, there is limited production in other areas, particularly in the central and southwestern portions of the state.

Fig. 43. Sugar Production as a Proportion of Total Acres of Farmland

Data Sources: U.S. Census Bureau; USDA, National Agricultural Statistics Service

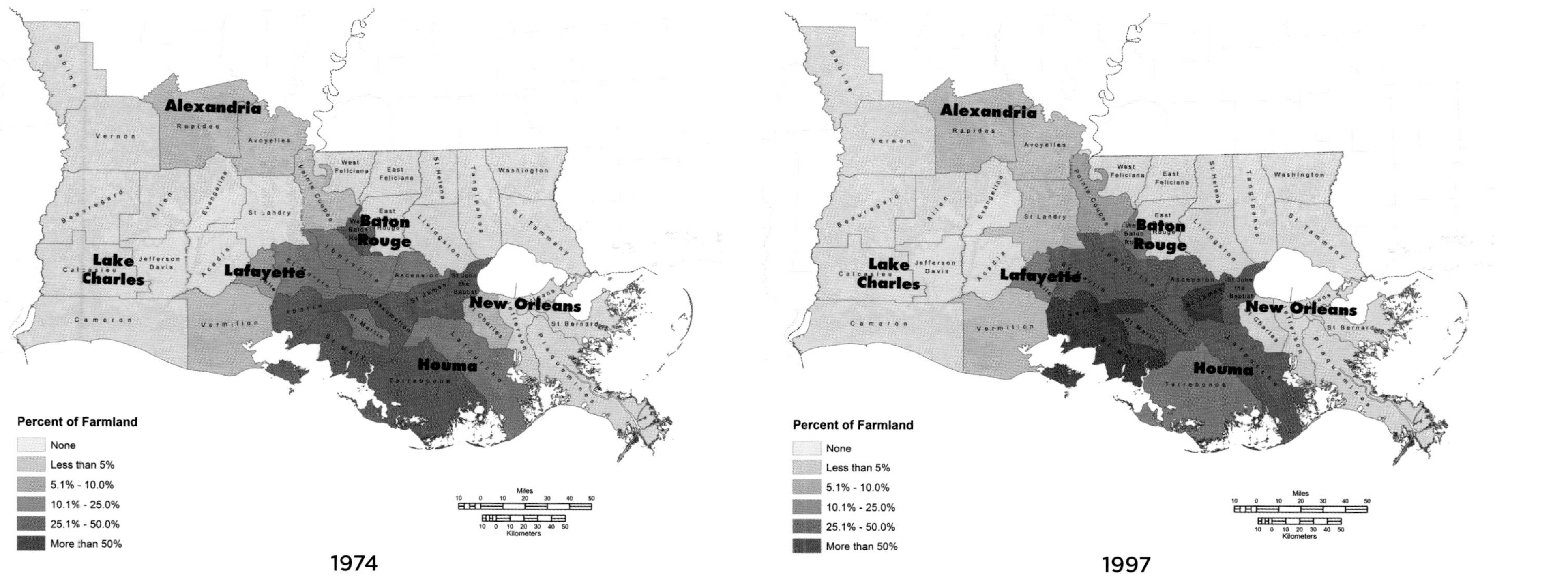

Fig. 44. Soy Production as a Proportion of Total Acres of Farmland

4 maps

Data Sources: U.S. Census Bureau; USDA, National Agricultural Statistics Service

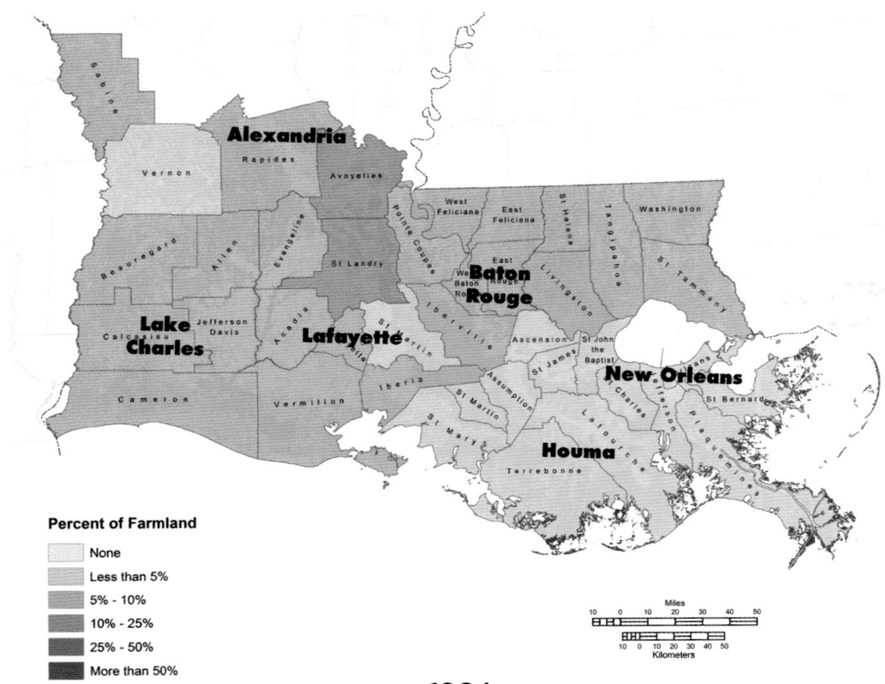

1964

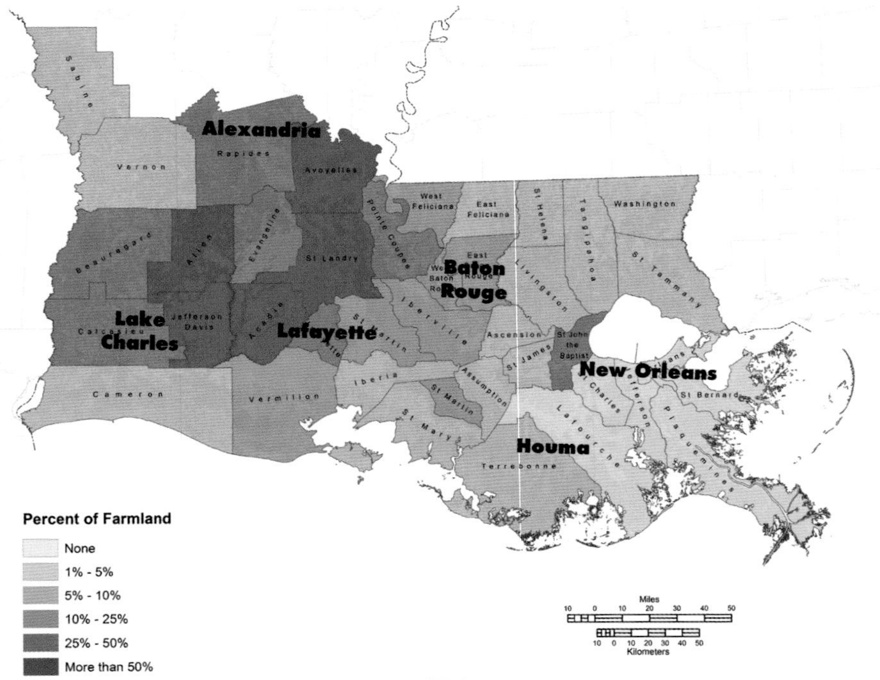

1974

New crops such as soybeans (fig. 44) have also been cultivated in response to changing global needs. Soybean farming rapidly expanded during the 1970s and 1980s as demand for soy oil and soy protein increased. Other cropland was converted to soybean fields, as were forests and pastureland. At first, soybean productivity was low due to pest and soil fertility problems, resulting in a brief contraction of the industry. As farming techniques improved, yields and profits climbed, allowing the crop to remain economically viable.

Fig. 44. Soy Production as a Proportion of Total Acres of Farmland

Data Sources: U.S. Census Bureau; USDA, National Agricultural Statistics Service

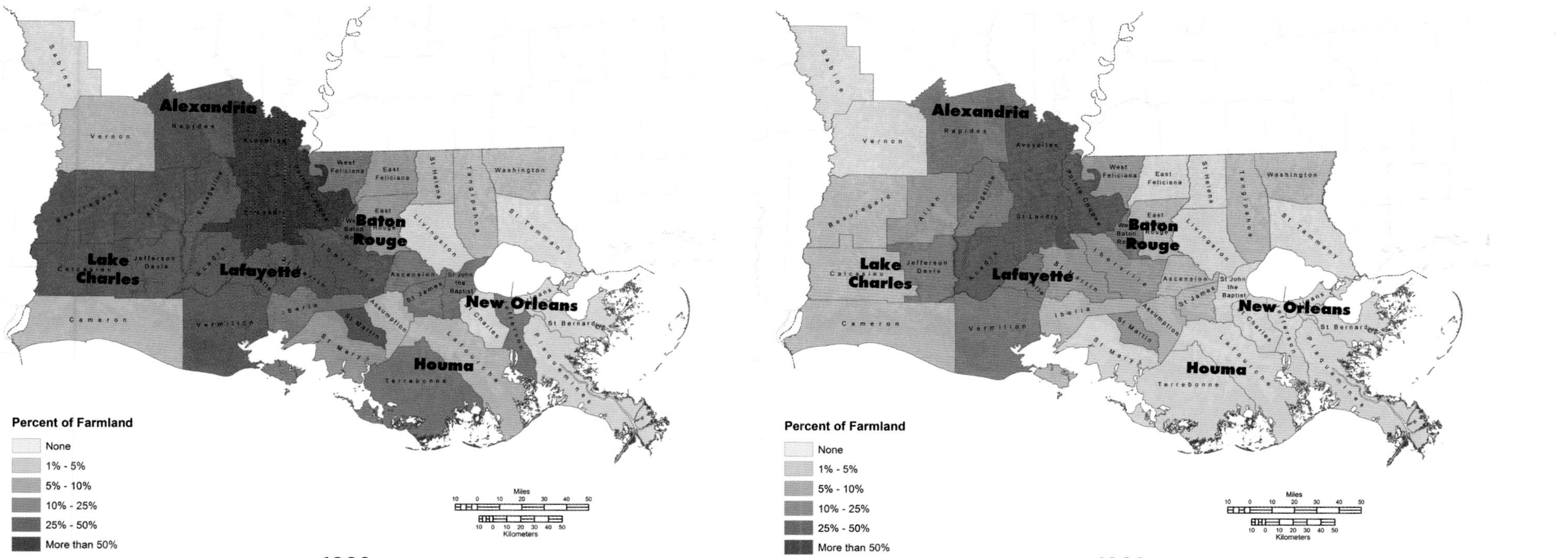

1982

1992

Fig. 45. Board Feet of Cypress Produced (3-Year Average)

3 maps

Data Sources: Louisiana Forestry Commission; Louisiana Department of Agriculture and Forestry

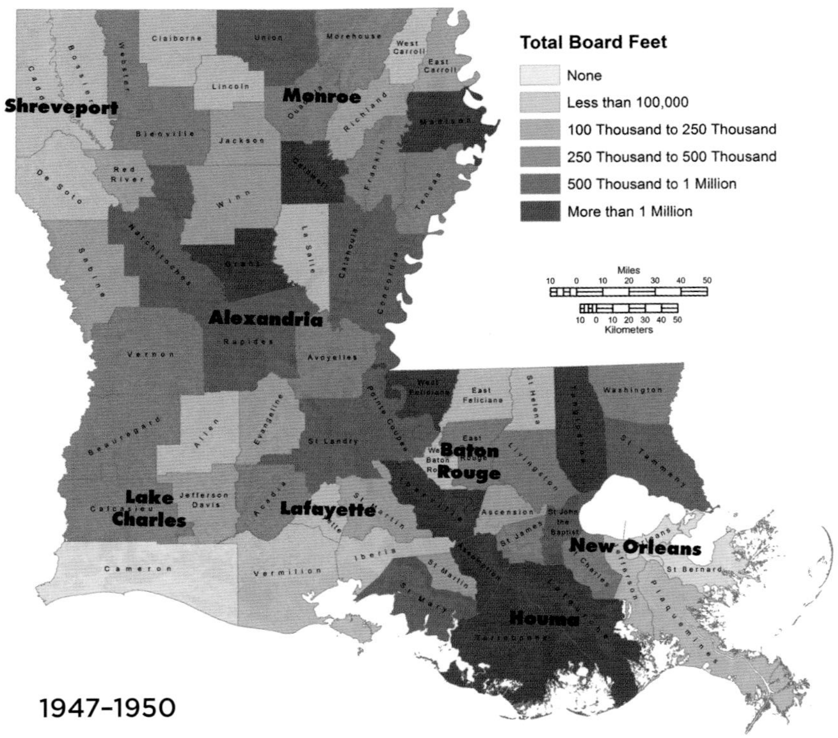

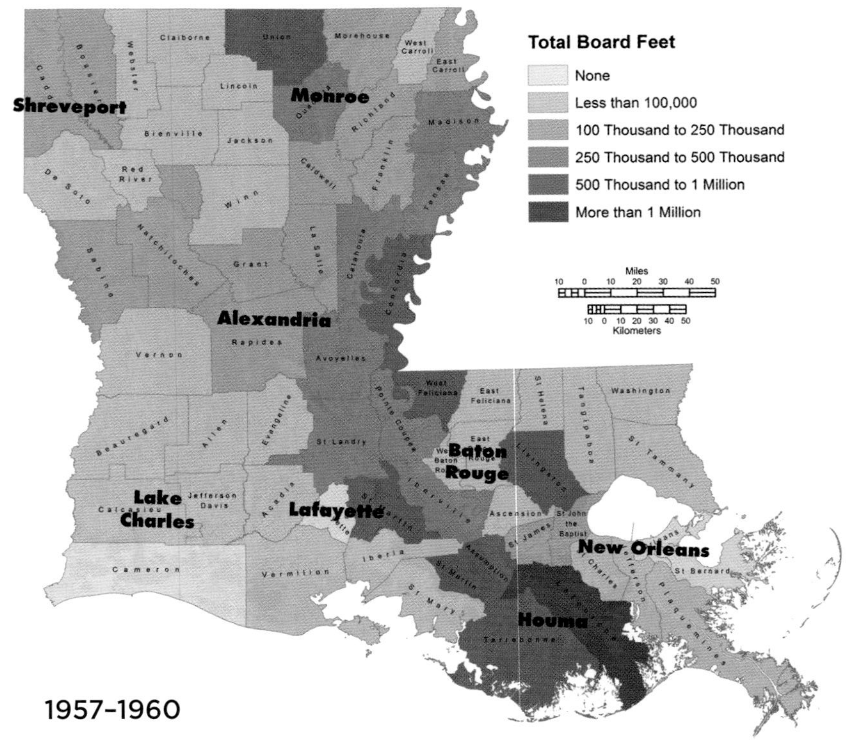

Once a dominant economic driver in the state, Louisiana's cypress industry peaked in the early years of the twentieth century and rapidly declined in the subsequent decades (fig. 45). A relatively small amount of cypress production continued in the state's river parishes in the latter half of the twentieth century, despite the closure of the last cypress mill in 1956.

Fig. 45. Board Feet of Cypress Produced (3-Year Average)

Data Sources: Louisiana Forestry Commission; Louisiana Department of Agriculture and Forestry

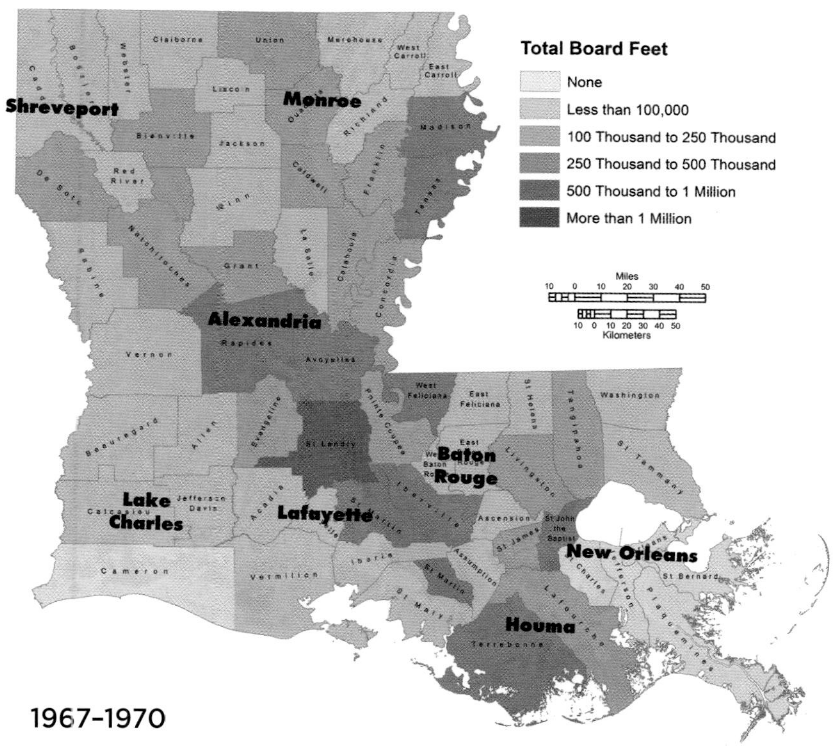

1967–1970

Fig. 46. Board Feet of Hardwood Produced (3-Year Average)

6 maps

Data Sources: Louisiana Forestry Commission; Louisiana Department of Agriculture and Forestry

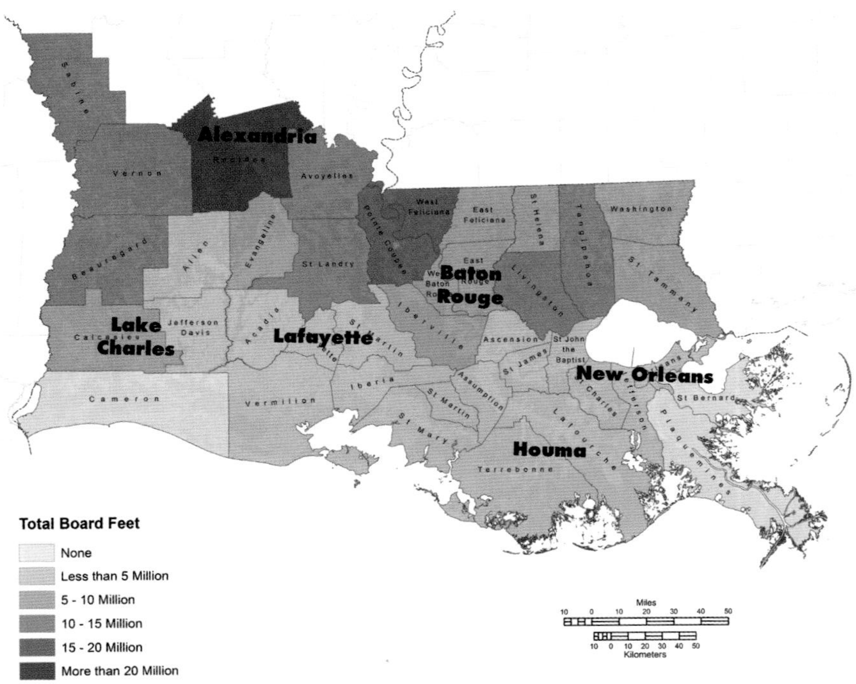

1947–1950

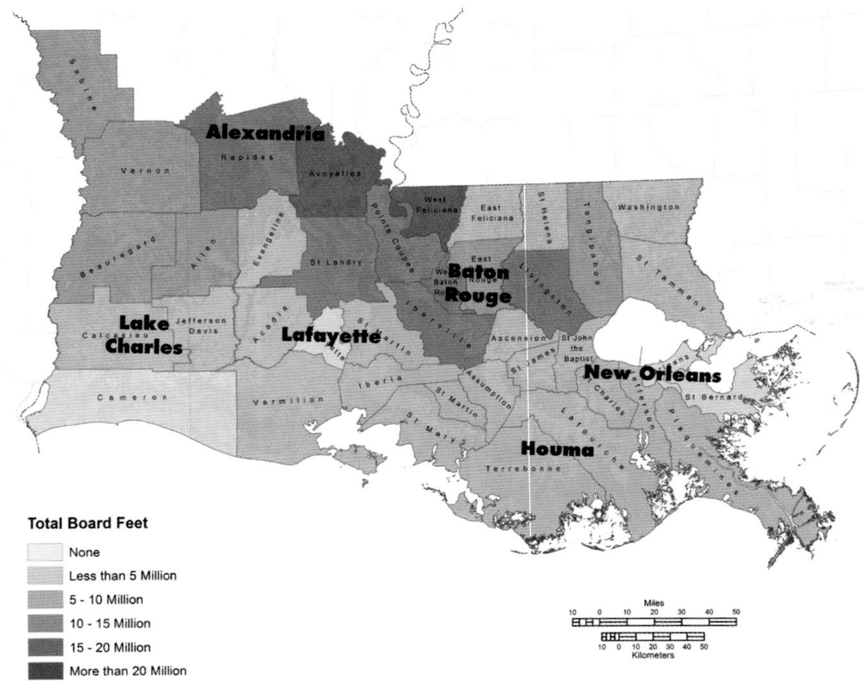

1957–1960

The same processes occurred in both the forestry and fishery industries. Following the collapse of Louisiana's cypress business, foresters shifted production to hardwoods and pine (figs. 46-47). While these industries had thrived in Louisiana long before the cypress boom, they became the primary focus of harvesting efforts once cypress stands had been depleted. Learning from the demise of cypress, Louisiana foresters developed effective conservation measures to keep the forestry industry strong.

Fig. 46. Board Feet of Hardwood Produced (3-Year Average)

Data Sources. Louisiana Forestry Commission; Louisiana Department of Agriculture and Forestry

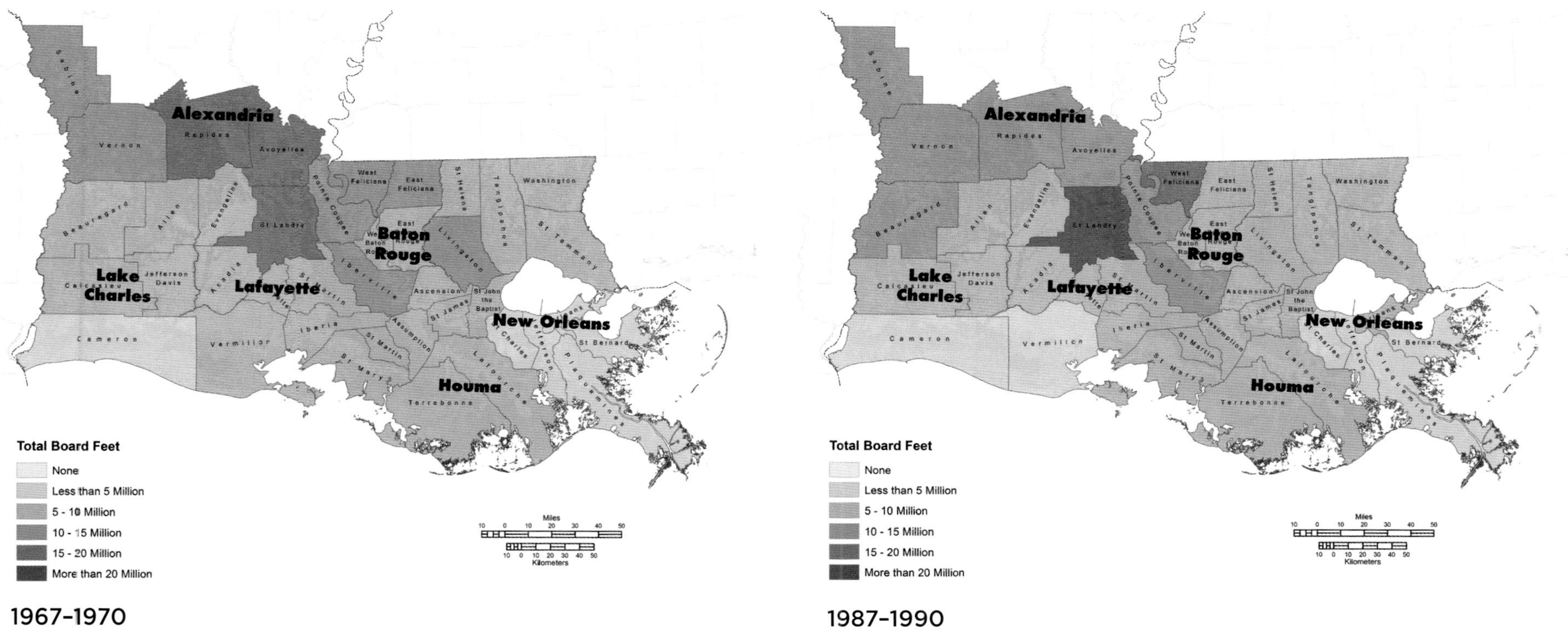

Fig. 46. Board Feet of Hardwood Produced (3-Year Average)

Data Sources: Louisiana Forestry Commission; Louisiana Department of Agriculture and Forestry

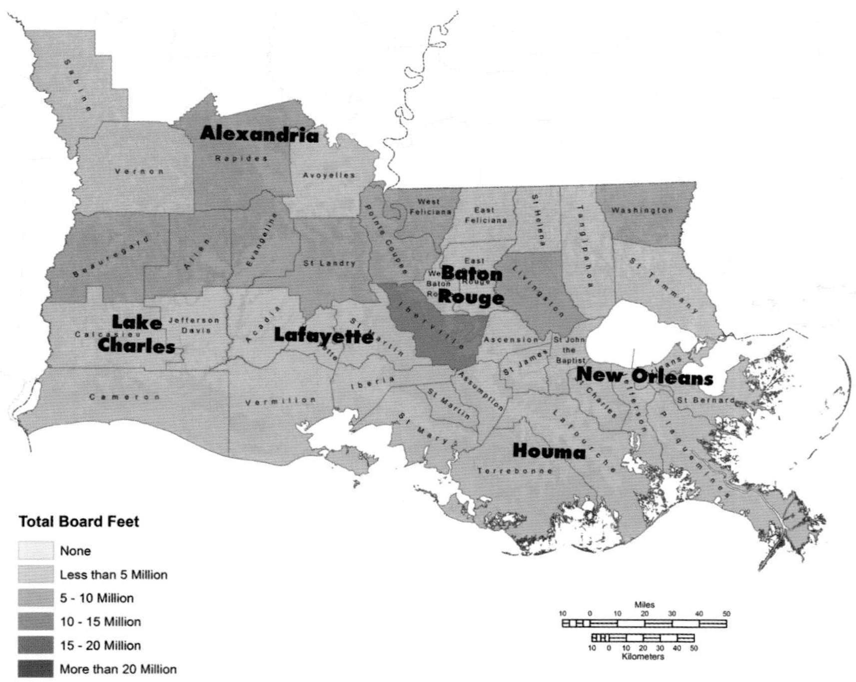

1997–2000

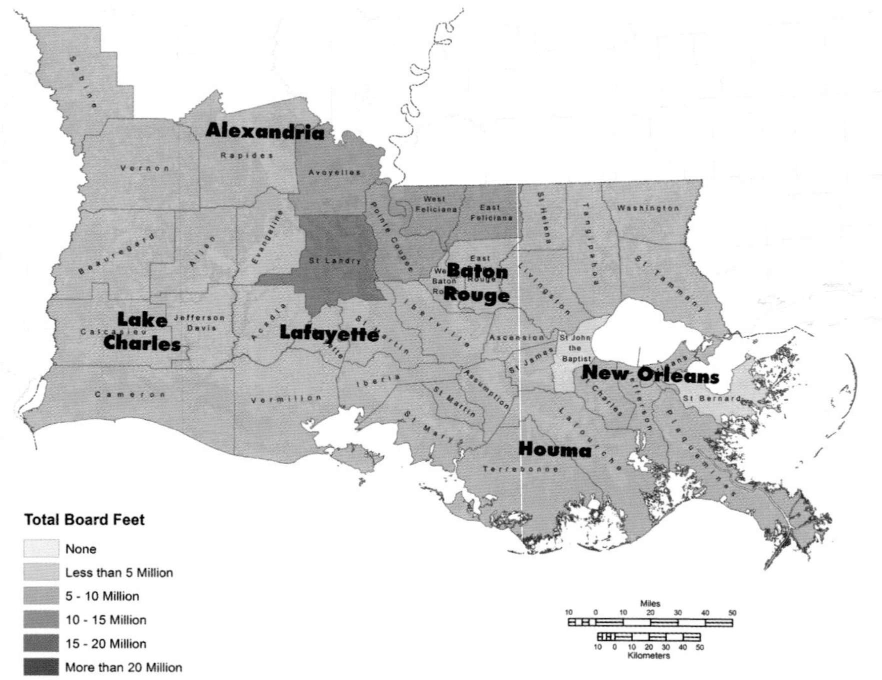

2007–2010

Fig. 47. Board Feet of Pine Produced (3-Year Average)

6 maps

Data Sources: Louisiana Forestry Commission; Louisiana Department of Agriculture and Forestry

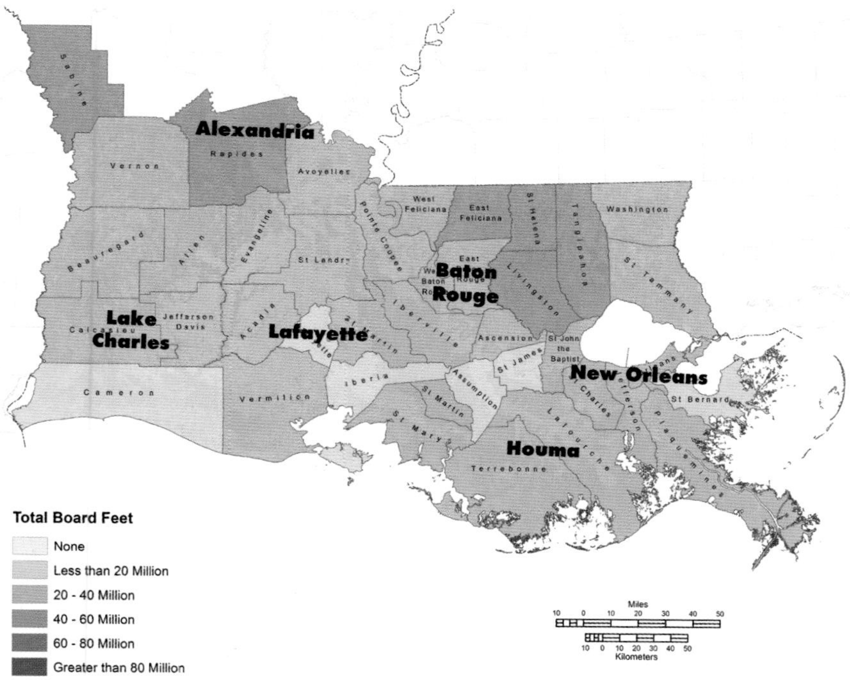

1947–1950

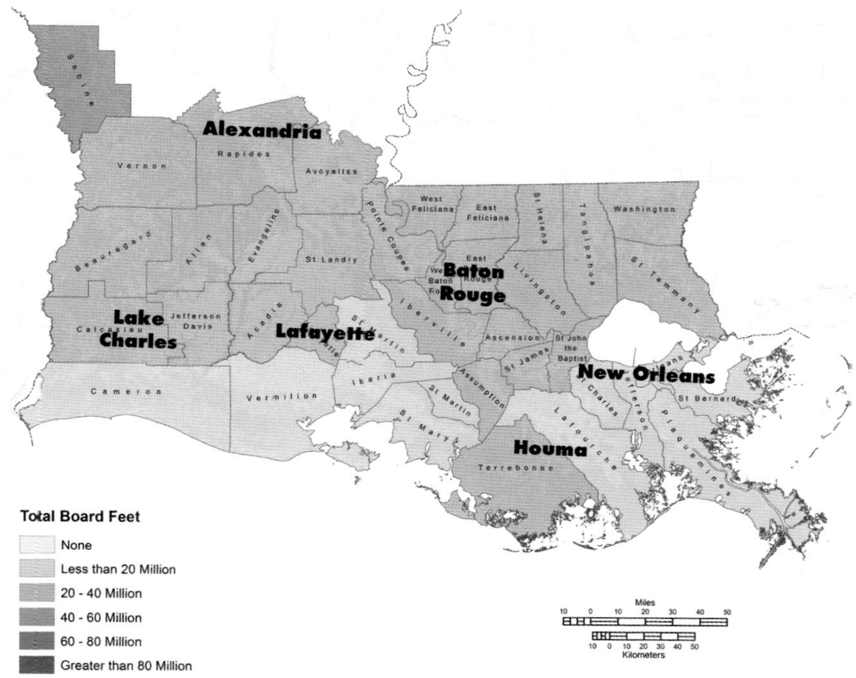

1957–1960

Fig. 47. Board Feet of Pine Produced (3-Year Average)

Data Sources: Louisiana Forestry Commission; Louisiana Department of Agriculture and Forestry

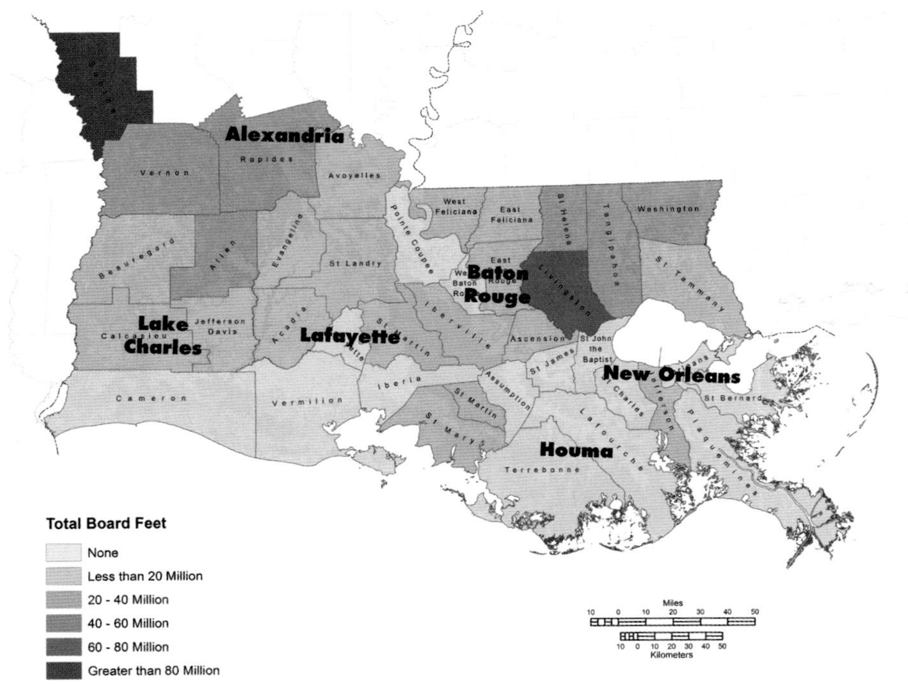

1967–1970

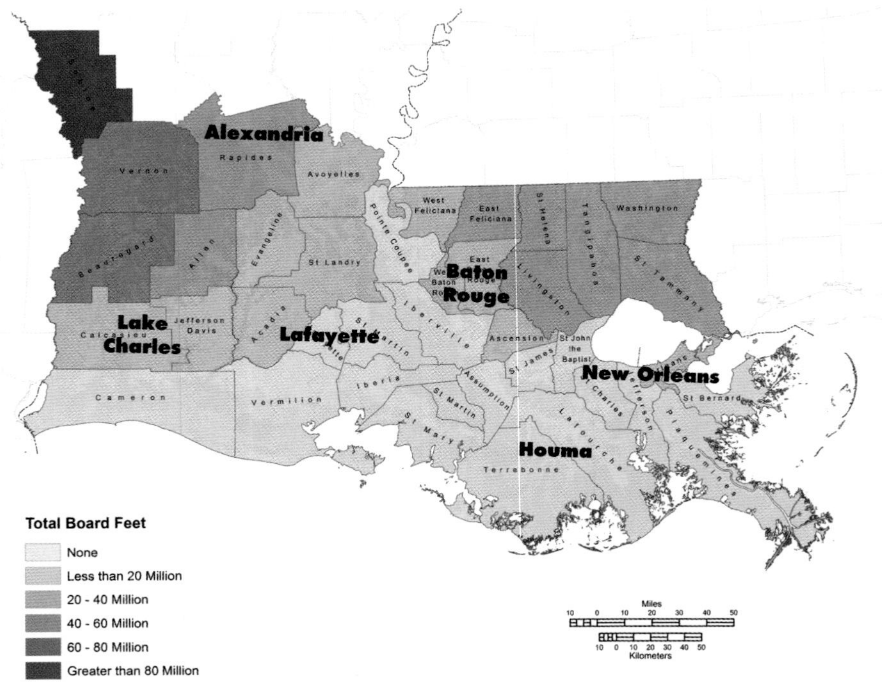

1987–1990

Fig. 47. Board Feet of Pine Produced (3-Year Average)

Data Sources: Louisiana Forestry Commission; Louisiana Department of Agriculture and Forestry

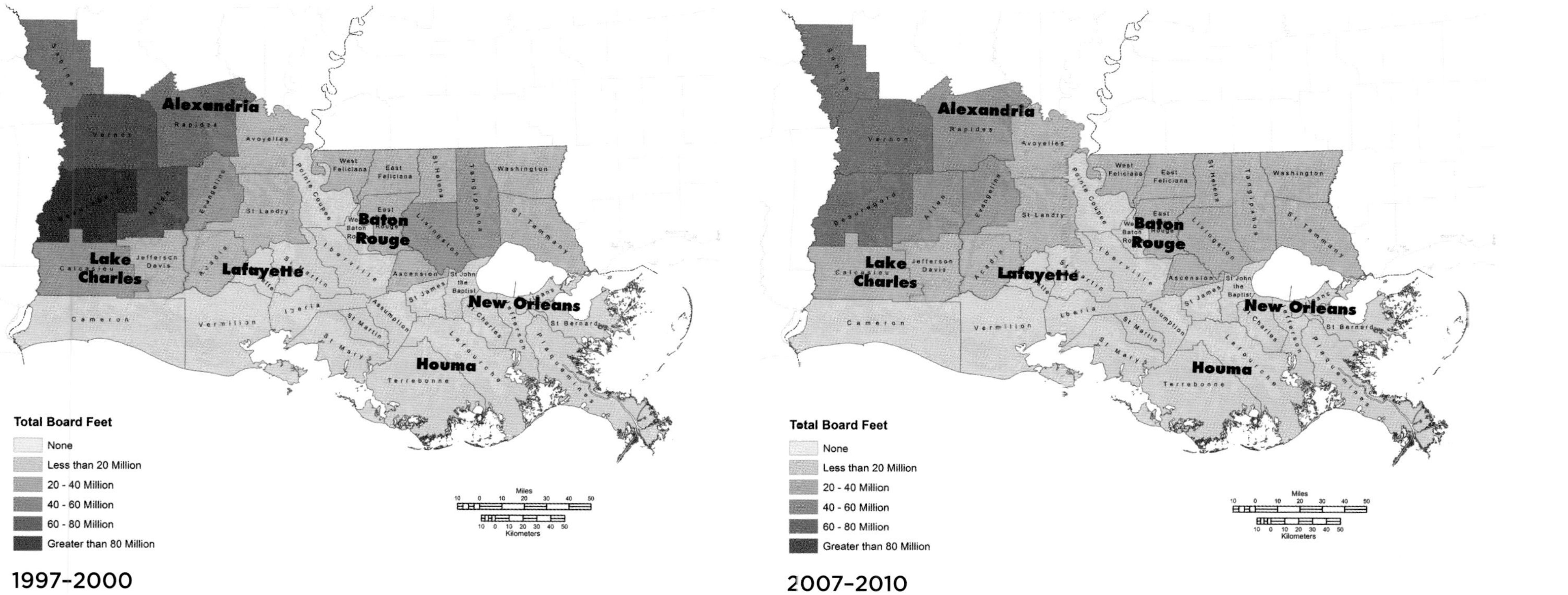

Fig. 48. Pounds of Oysters Landed

6 maps

Data Sources: U.S. Department of the Interior, Bureau of Commercial Fisheries; NOAA National Marine Fisheries Service

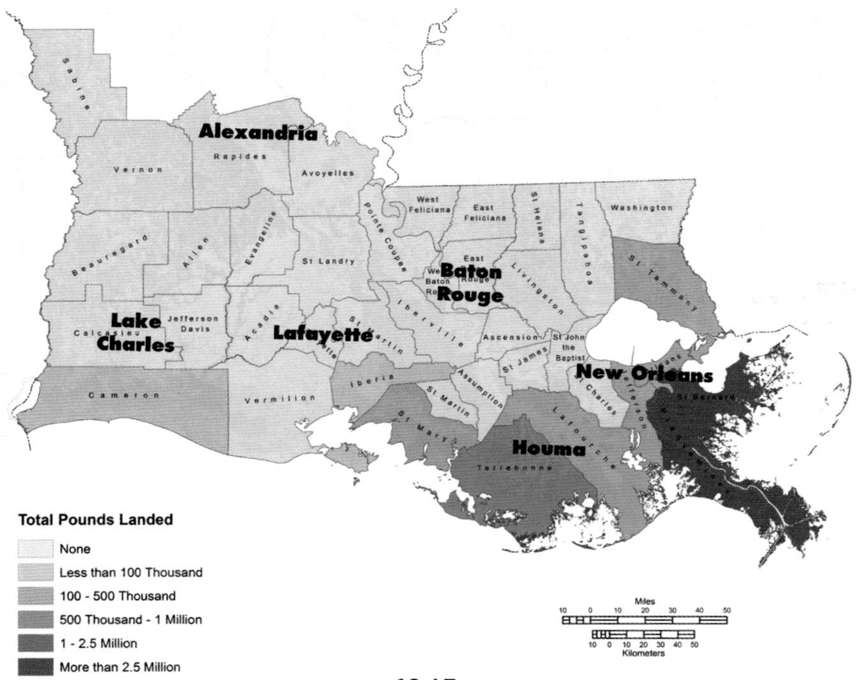

1945

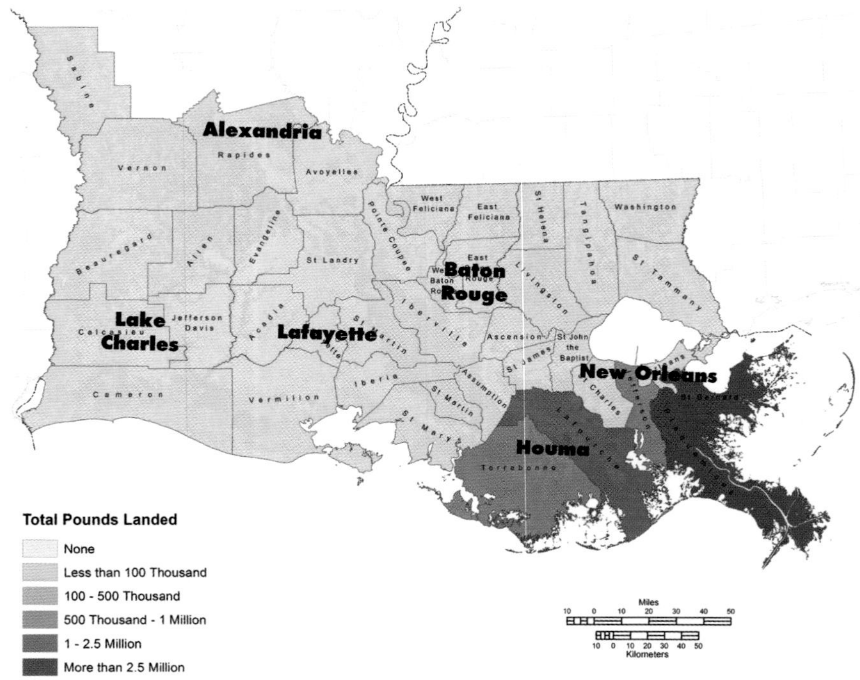

1955

While fluctuations in fisheries landings have occurred over time, oysters and shrimp remained important landings throughout the latter half of the twentieth century. Louisiana's oyster industry is one of the most successful fisheries in the United States due in large part to the effective management and cultivation of the seed grounds (fig. 48). Oyster production in the state has historically revolved around the seed ground of southeast Louisiana, from St. Bernard Parish to Terrebonne Parish. Recent decades have seen an increasing development of the oyster industry in some southwestern parishes, particularly Cameron and Vermilion.

Fig. 48. Pounds of Oysters Landed

Data Sources: U.S. Department of the Interior, Bureau of Commercial Fisheries; NOAA National Marine Fisheries Service

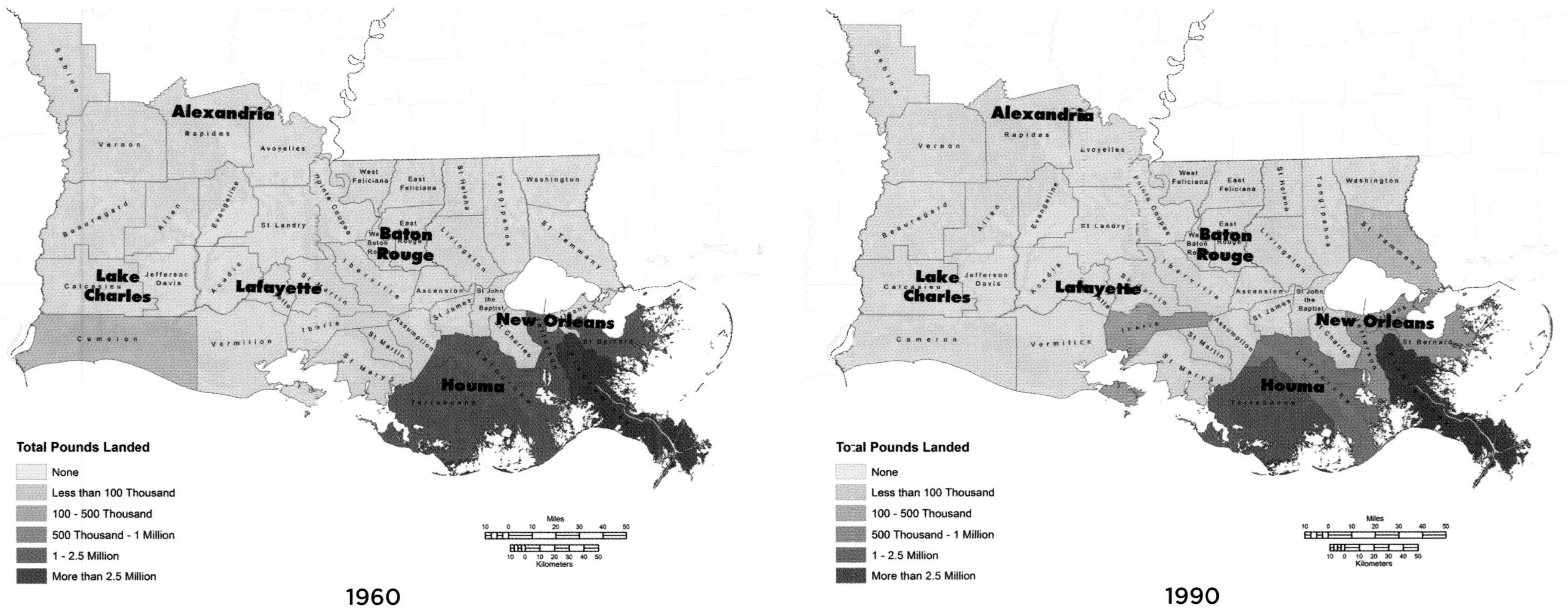

Fig. 48. Pounds of Oysters Landed

Data Sources: U.S. Department of the Interior, Bureau of Commercial Fisheries; NOAA National Marine Fisheries Service

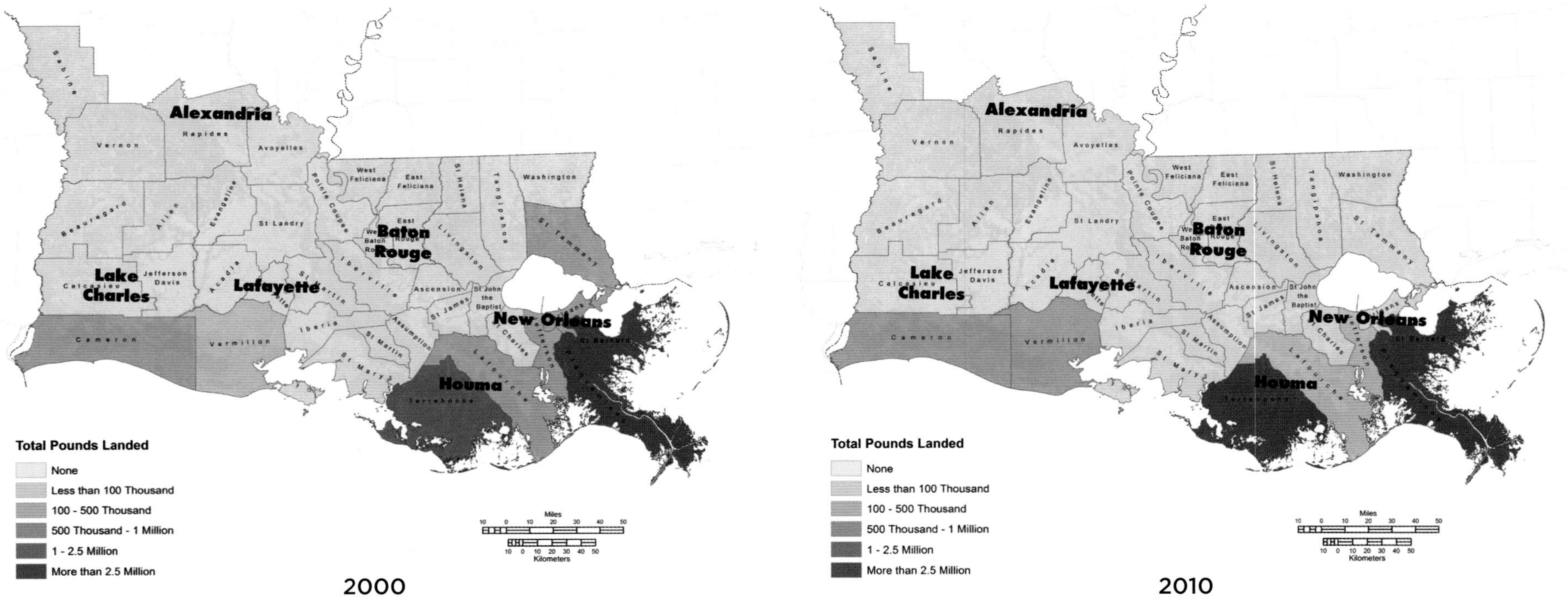

Fig. 49. Pounds of Shrimp Landed

6 maps

Data Sources: U.S. Department of the Interior, Bureau of Commercial Fisheries; NOAA National Marine Fisheries Service

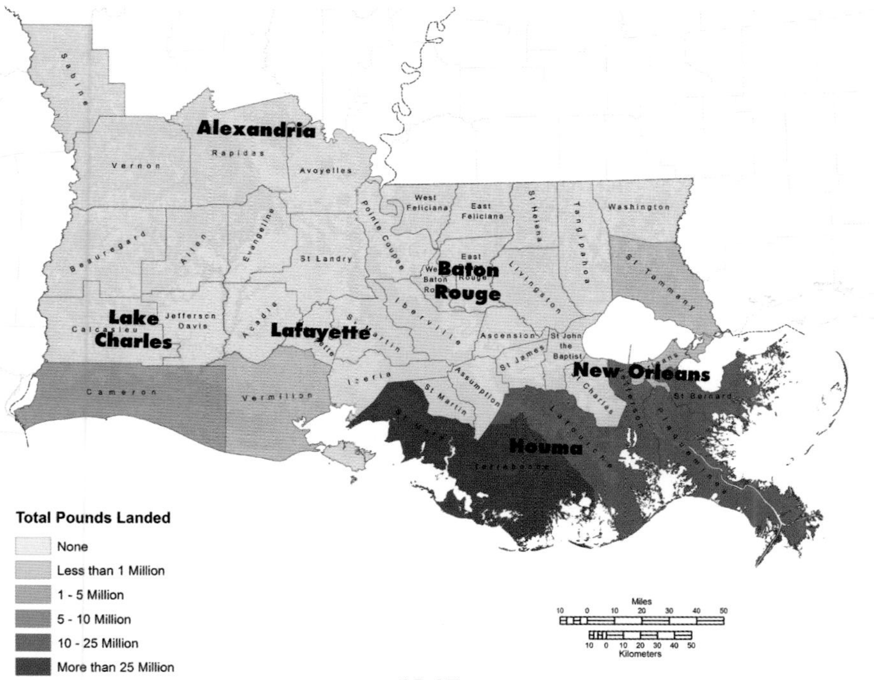

1945

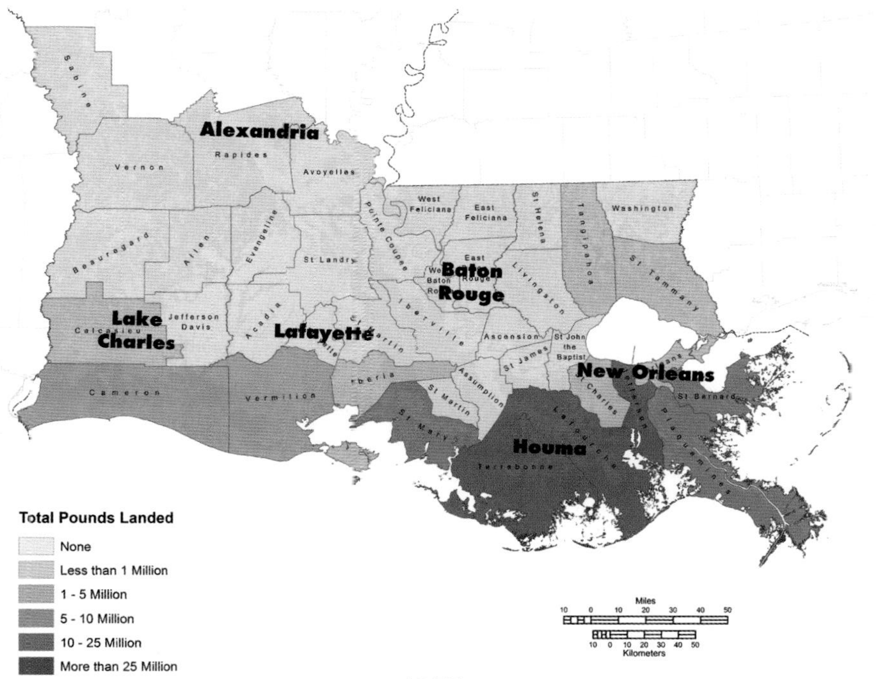

1955

The shrimp industry is one of the most widespread coastal fisheries in the state, with production occurring across all coastal parishes (fig. 49). The shrimp industry began in the state's inland waterways and shallow bays, using smaller vessels that could operate in shallow waters. This shallow-water shrimping industry continues to this day. In the latter half of the twentieth century, however, with the development of steel-hulled vessels and refrigeration, some fishermen began to work in the offshore waters of the Gulf of Mexico. Despite some fluctuations, particularly along the central coast, the pounds of shrimp landed have remained at a steadily high level since the 1940s.

Fig. 49. Pounds of Shrimp Landed

Data Sources: U.S. Department of the Interior, Bureau of Commercial Fisheries; NOAA National Marine Fisheries Service

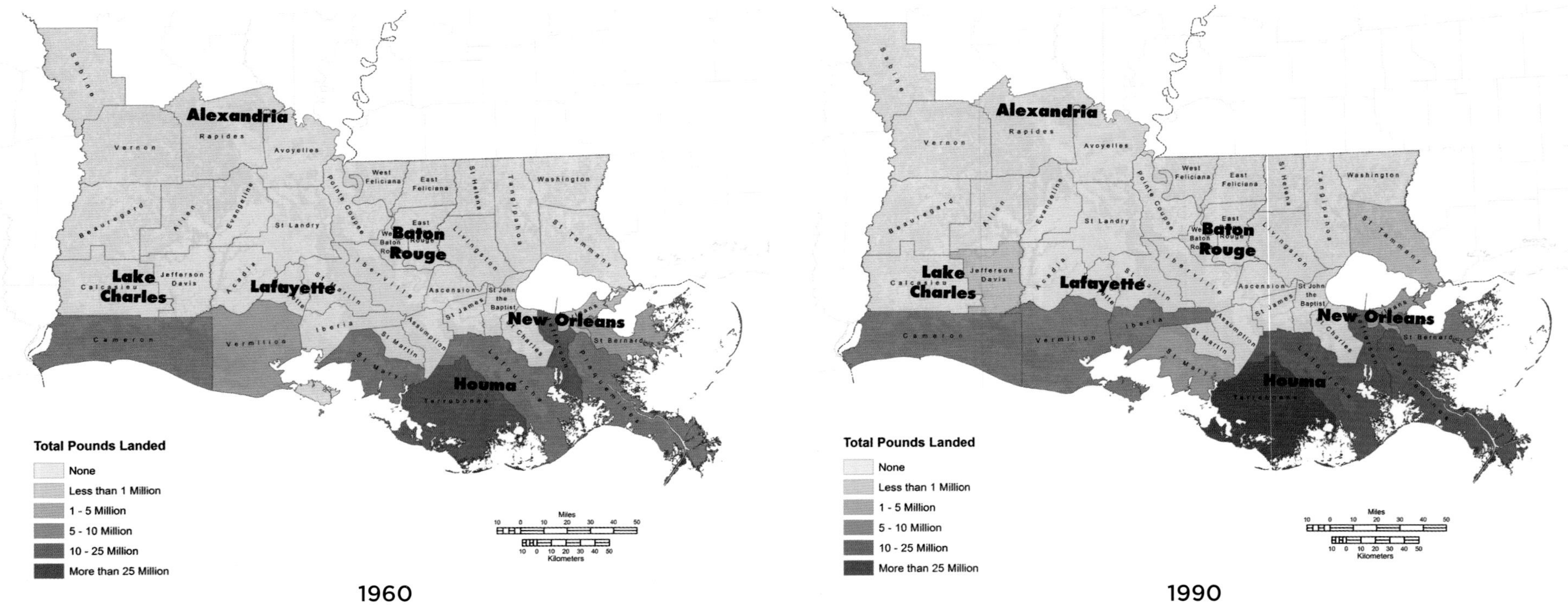

Fig. 49. Pounds of Shrimp Landed

Data Sources: U.S. Department of the Interior, Bureau of Commercial Fisheries; NOAA National Marine Fisheries Service

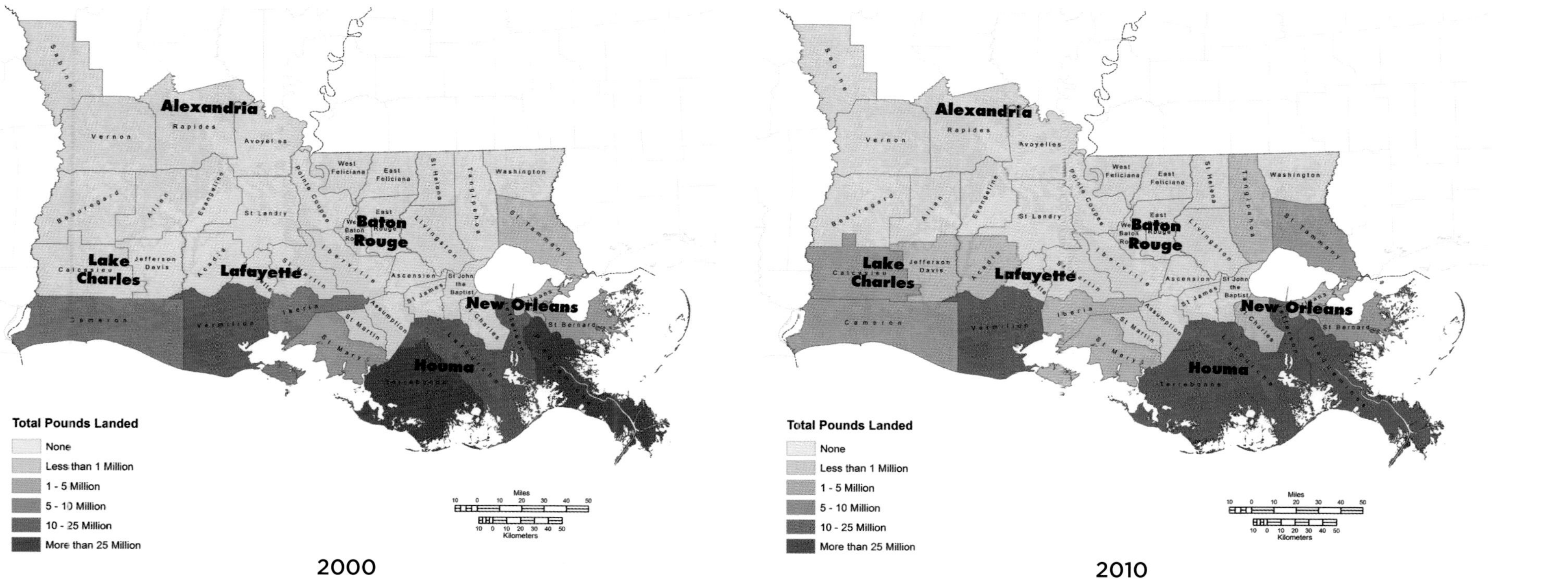

Fig. 50. Pounds of Menhaden Landed

6 maps

Data Sources: U.S. Department of the Interior, Bureau of Commercial Fisheries; NOAA National Marine Fisheries Service

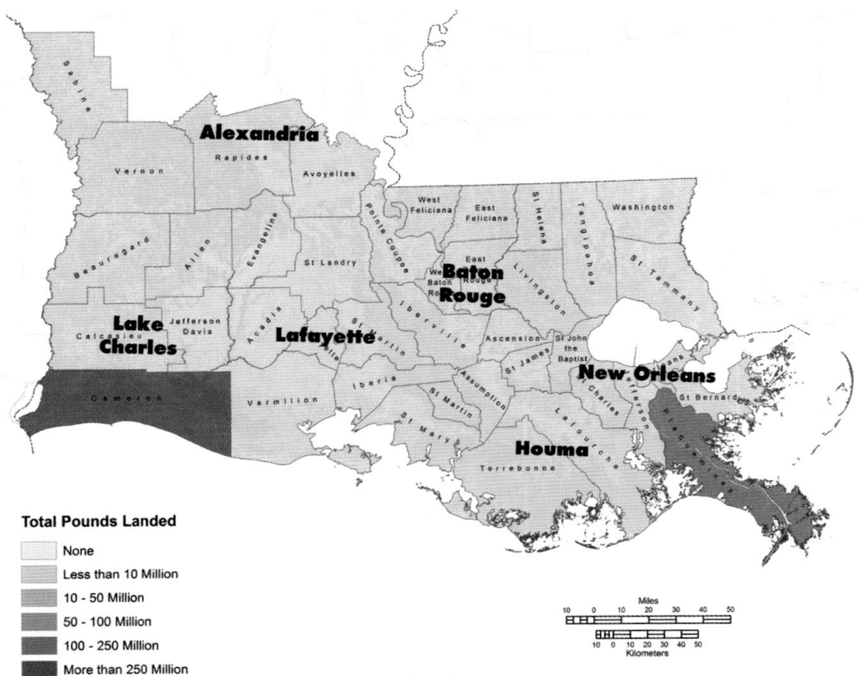

1950

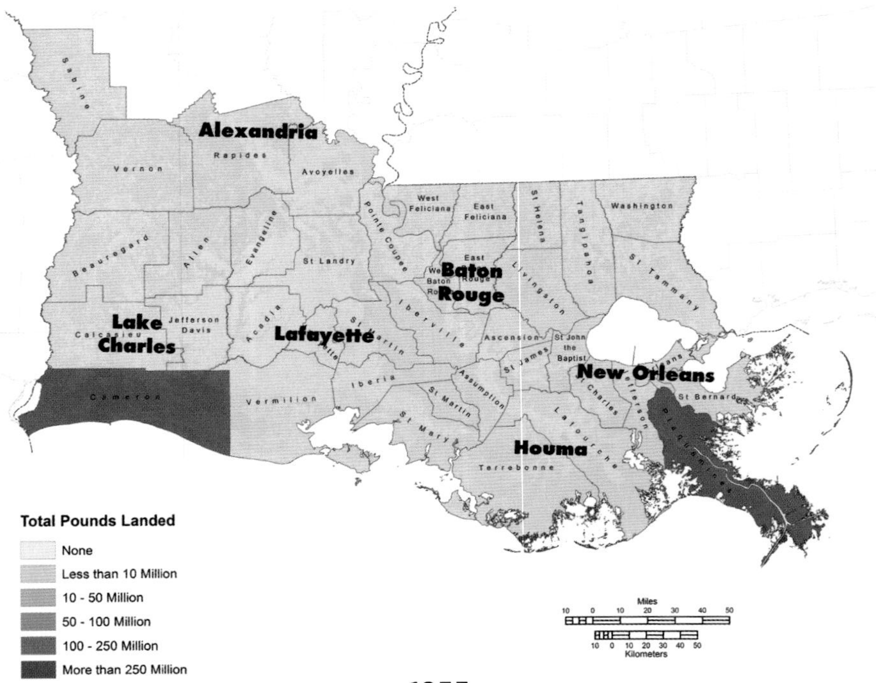

1955

As with the new crop of soybeans in agriculture, Louisiana's fisheries experienced the relatively recent growth and development of a new resource in the latter half of the twentieth century, the Gulf menhaden (fig. 50). This fish is generally considered unfit for human consumption due to its high oil content. However, these small fish are ground into fish meal and oil for use in pet feed and various industrial products. The development of improved technology during the 1950s made the landing of these small fish much more economical. As a result, a number of menhaden-processing plants were established in coastal Louisiana, diversifying the state's commercial fisheries. Most notable locations in coastal Louisiana are Cameron and Plaquemines parishes.

Fig. 50. Pounds of Menhaden Landed

Data Sources: U.S. Department of the Interior, Bureau of Commercial Fisheries; NOAA National Marine Fisheries Service

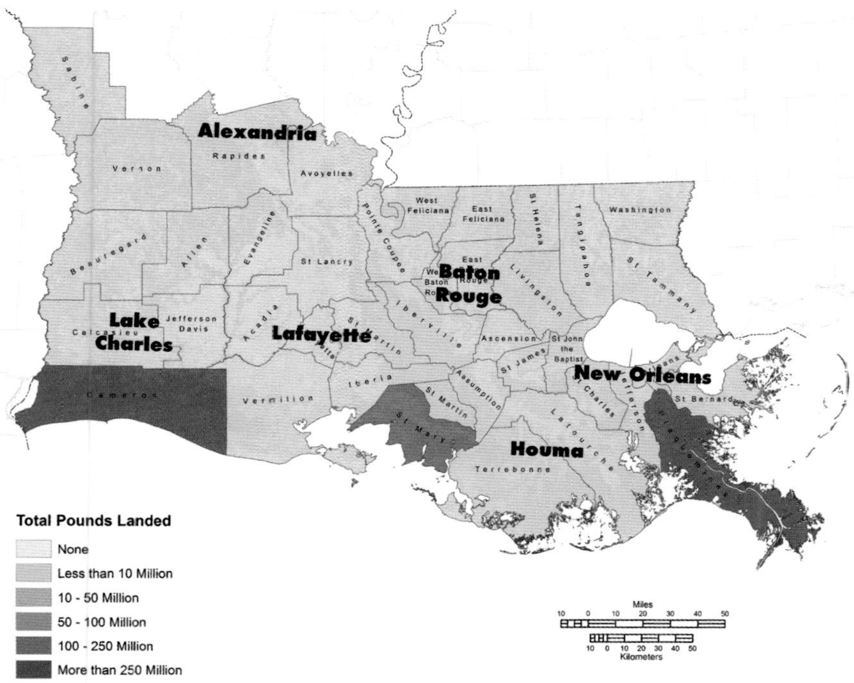

1960

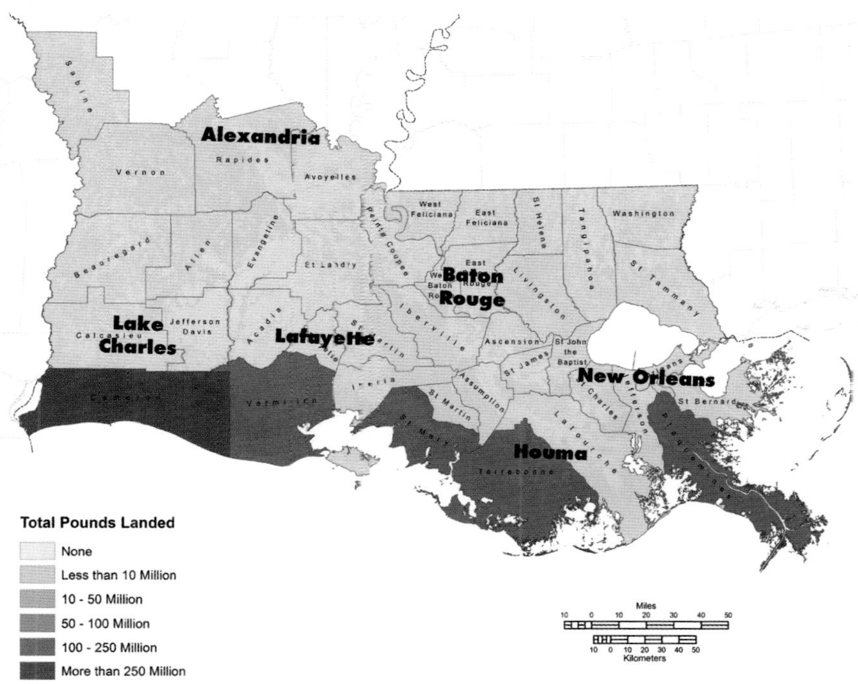

1990

Fig. 50. Pounds of Menhaden Landed

Data Sources: U.S. Department of the Interior, Bureau of Commercial Fisheries; NOAA National Marine Fisheries Service

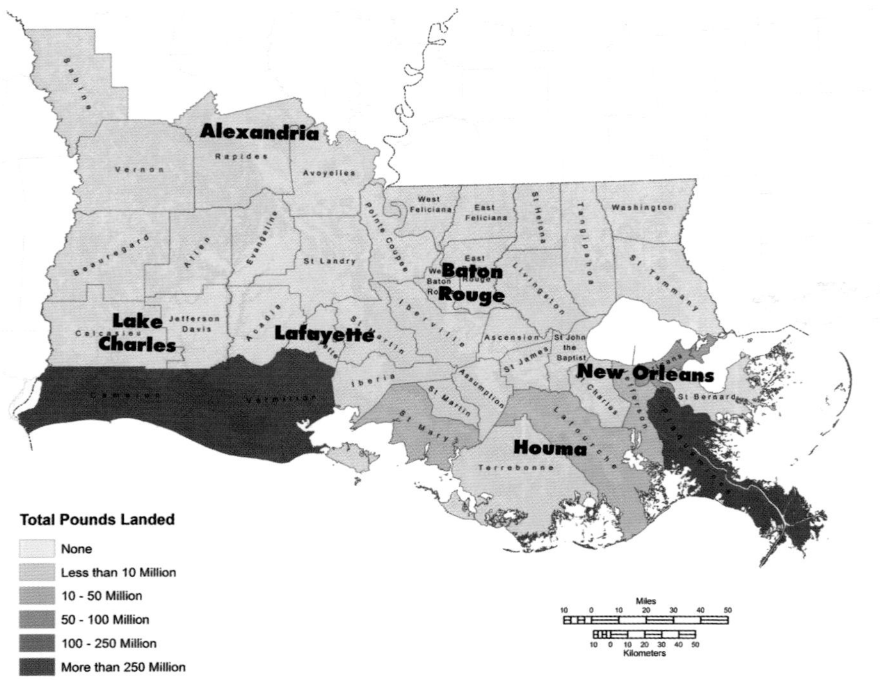

2000

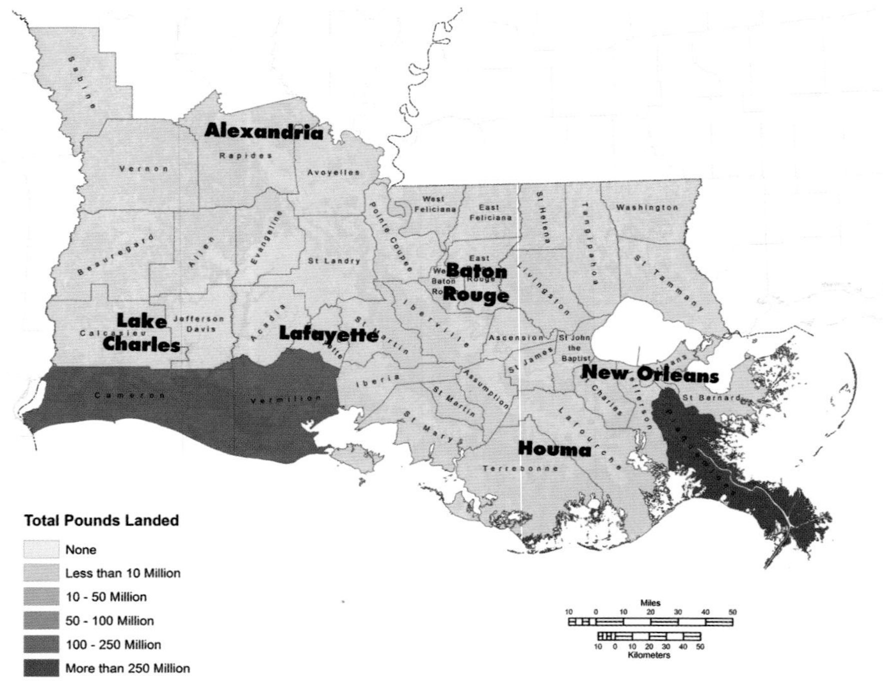

2010

Fig. 51. Cumulative Density of Onshore Gas Wells (1950–2010)

7 maps

Data Source: Louisiana Department of Natural Resources

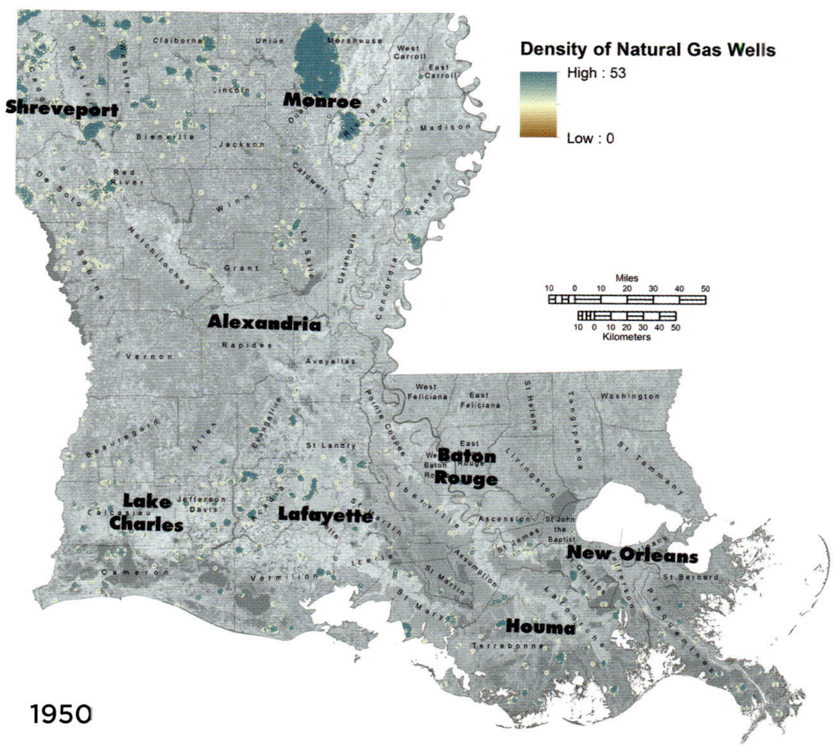

1950

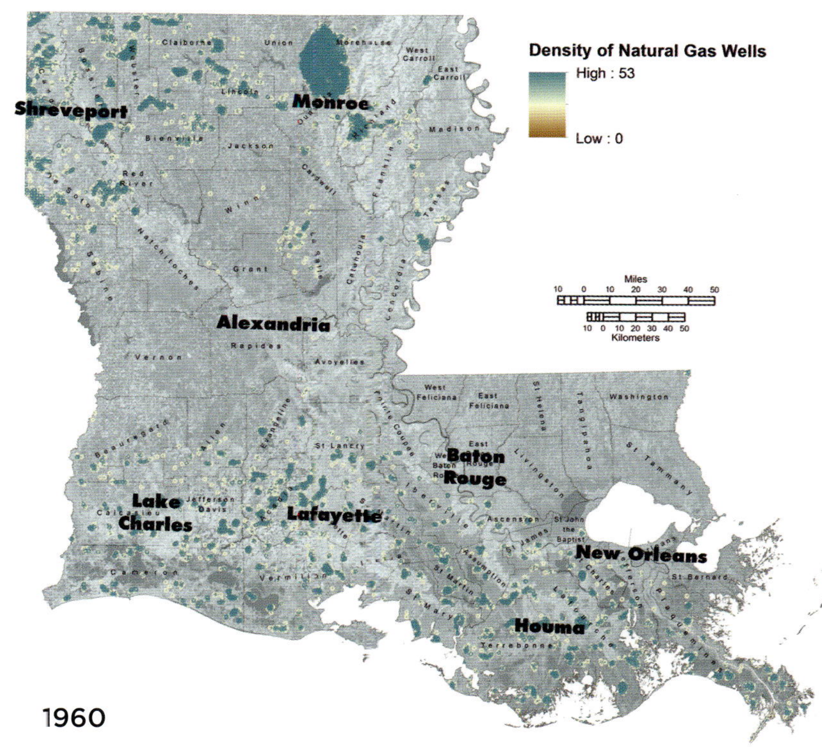

1960

The development of new technology has been a driving force behind many coastal innovations and adaptations. This has been especially true of the oil and gas industry in Louisiana. Natural gas was first discovered in Louisiana in 1870 when drillers were searching for artesian water sources, and the first natural gas pipeline in Louisiana was laid in 1908. The industry would continue to grow until 1970, when the natural gas reserves dropped for the first time (fig. 51). In 2008, the Haynesville Shale in northwestern Louisiana became a major producer of shale gas, leading to a continuing expansion of natural gas wells in this area.

Fig. 51. Cumulative Density of Onshore Gas Wells (1950–2010)

Data Source: Louisiana Department of Natural Resources

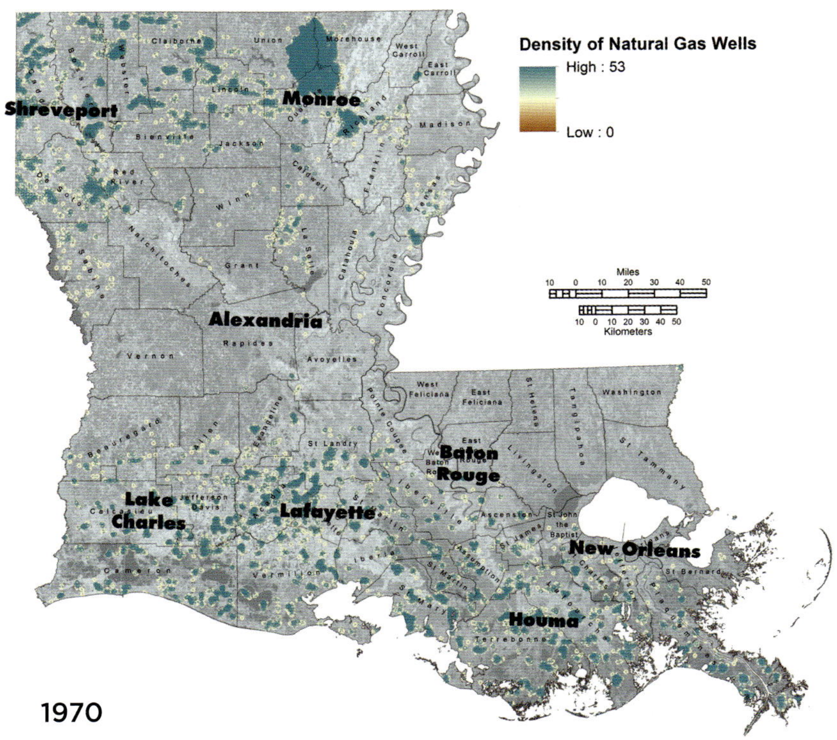

1970

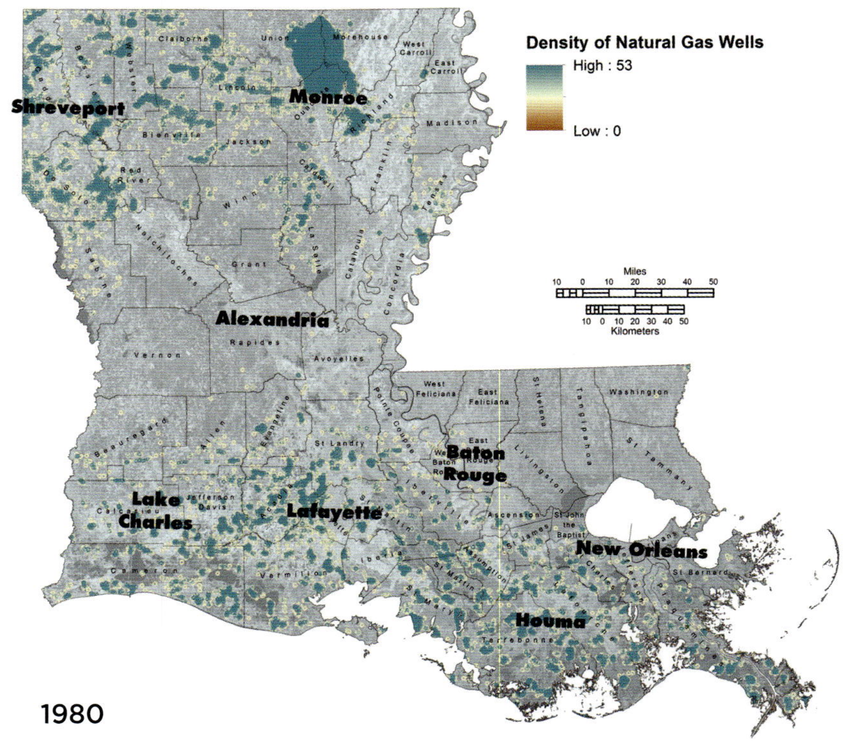

1980

Fig. 51. Cumulative Density of Onshore Gas Wells (1950–2010)

Data Source: Louisiana Department of Natural Resources

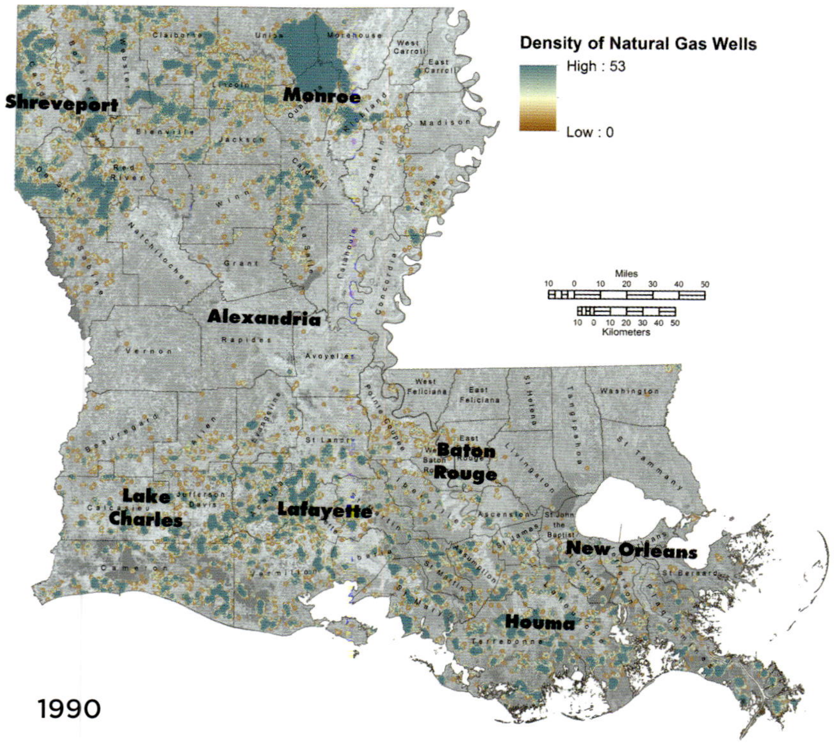

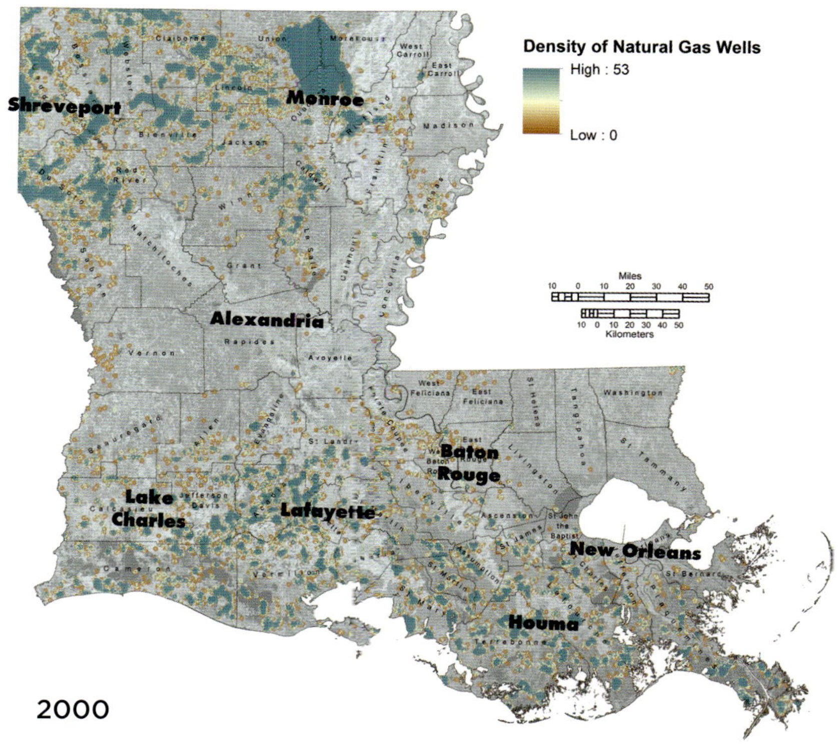

Fig. 51. Cumulative Density of Onshore Gas Wells (1950–2010)

Data Source: Louisiana Department of Natural Resources

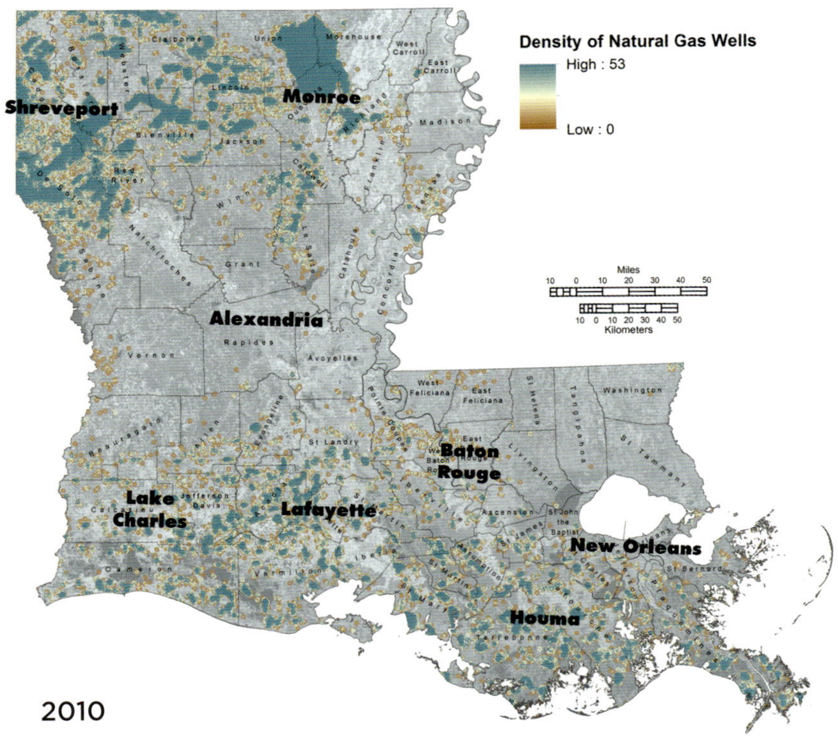

2010

Fig. 52. Cumulative Density of Onshore Oil Wells (1950–2010)

7 maps

Data Source: Louisiana Department of Natural Resources

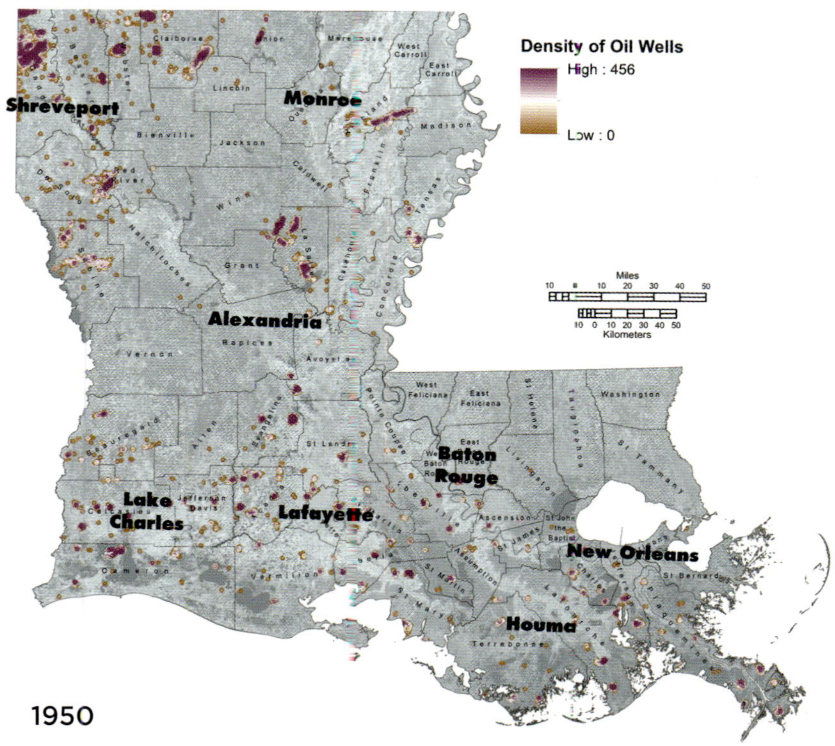

1950

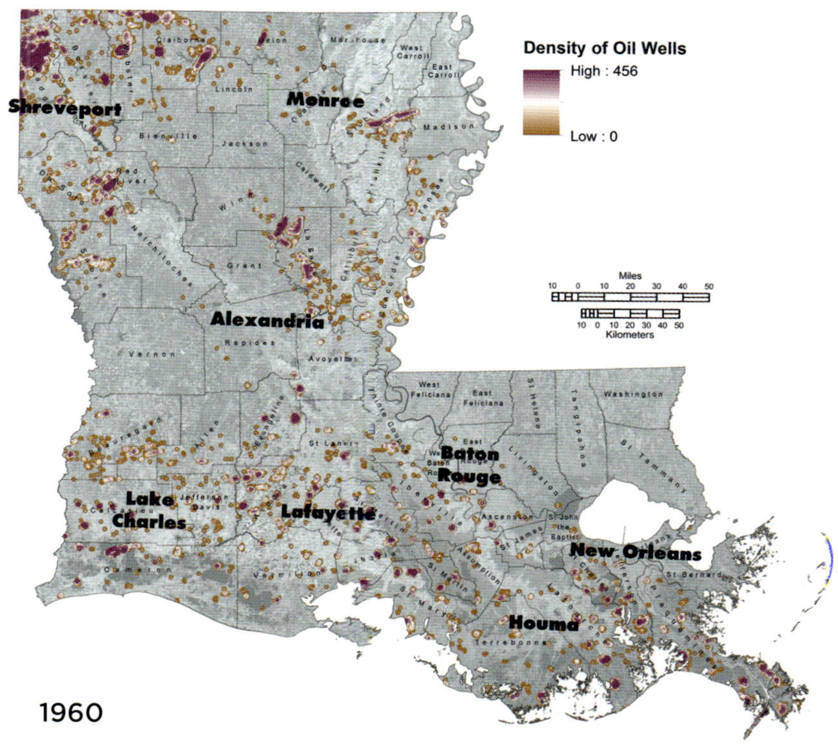

1960

The first active oil wells in Louisiana were drilled in 1901 in the town of Jennings, located in Jefferson Davis Parish in southwest Louisiana, and in 1904 in Caddo Parish, in the northern tier of the state. Following these discoveries, oil production boomed in both the northern and southern portions of the state, with significantly little development occurring in central Louisiana (fig. 52). Production would continue to climb until reaching a peak in 1969, after which the state's onshore oil reserves began to decline.

Fig. 52. Cumulative Density of Onshore Oil Wells (1950–2010)

Data Source: Louisiana Department of Natural Resources

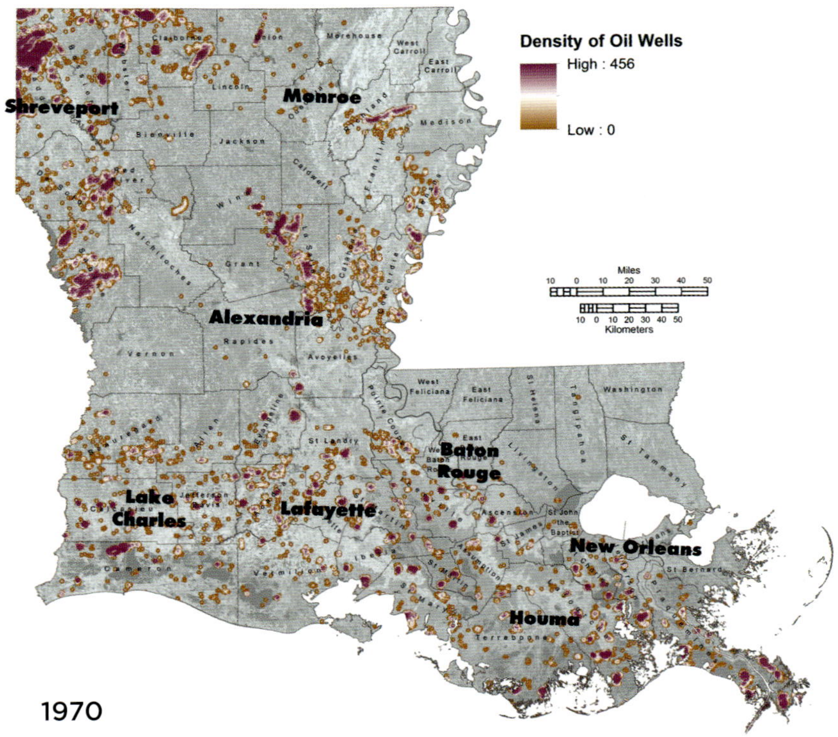

1970

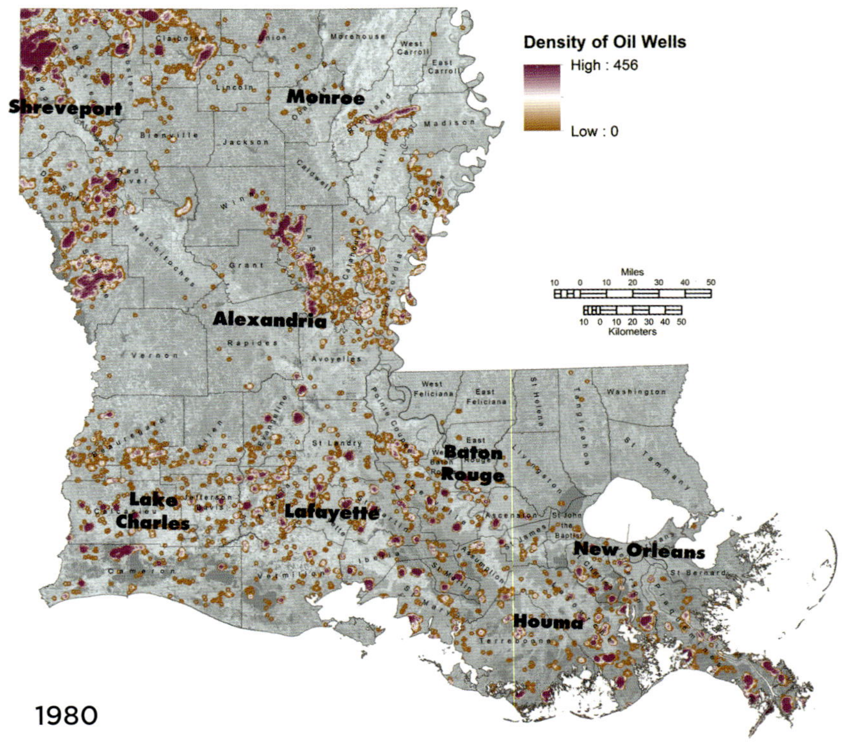

1980

Fig. 52. Cumulative Density of Onshore Oil Wells (1950–2010)

Data Source: Louisiana Department of Natural Resources

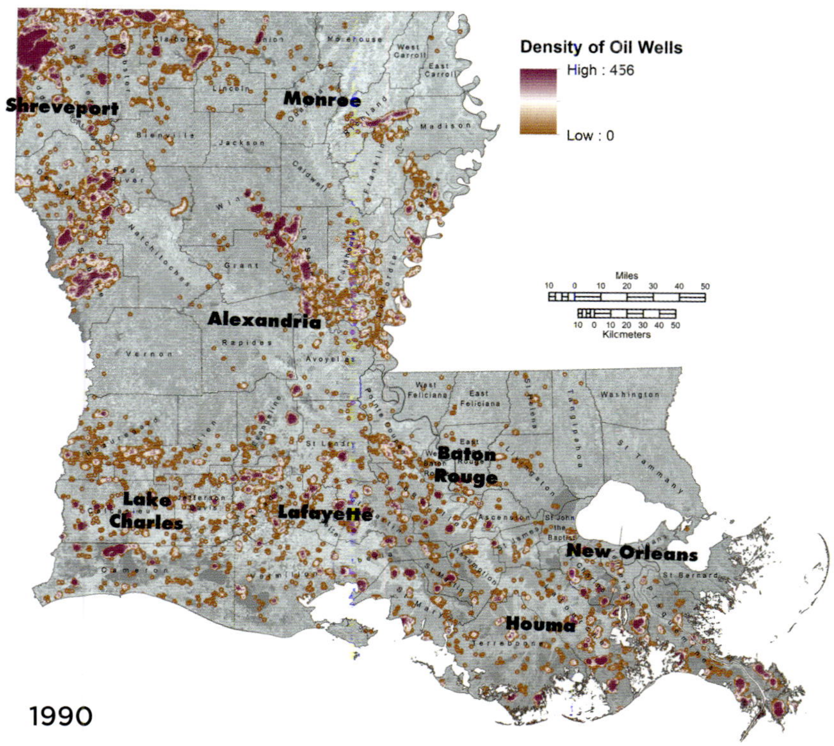

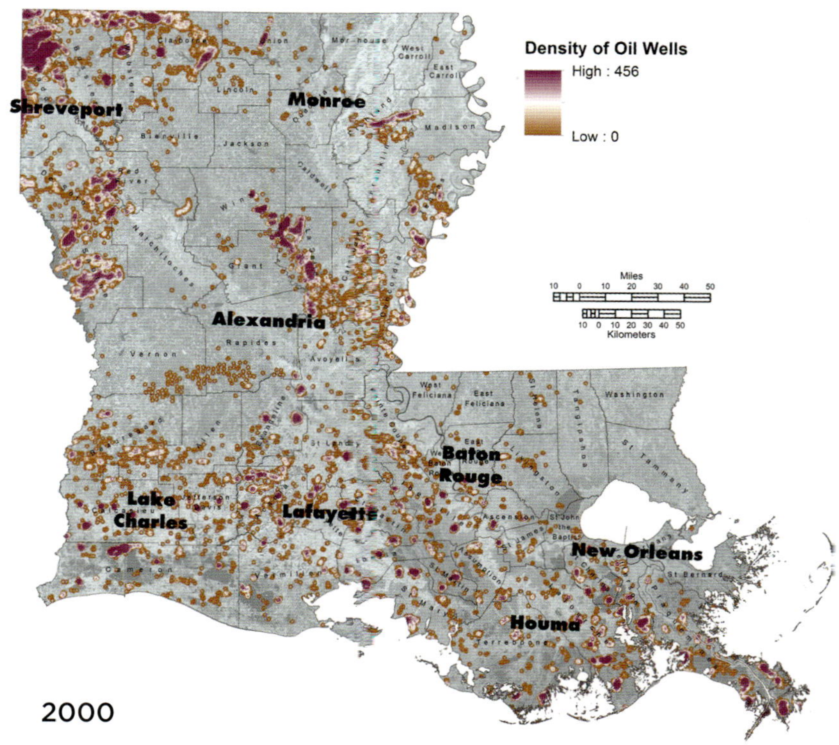

Fig. 52. Cumulative Density of Onshore Oil Wells (1950–2010)

Data Source: Louisiana Department of Natural Resources

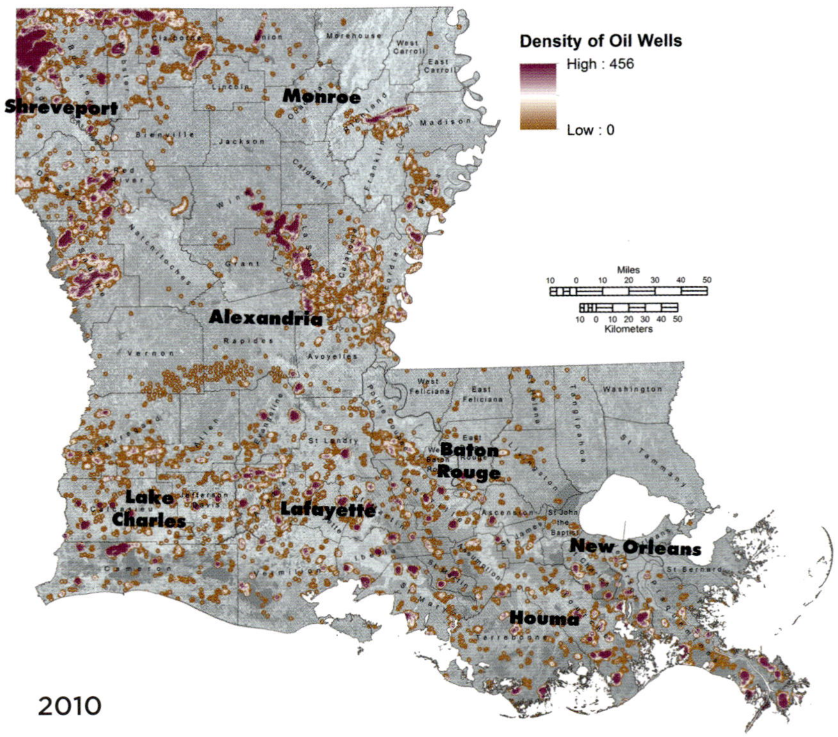

2010

Fig. 53. Density of Nearshore Gas Wells (2010)

Data Source: Louisiana Department of Natural Resources

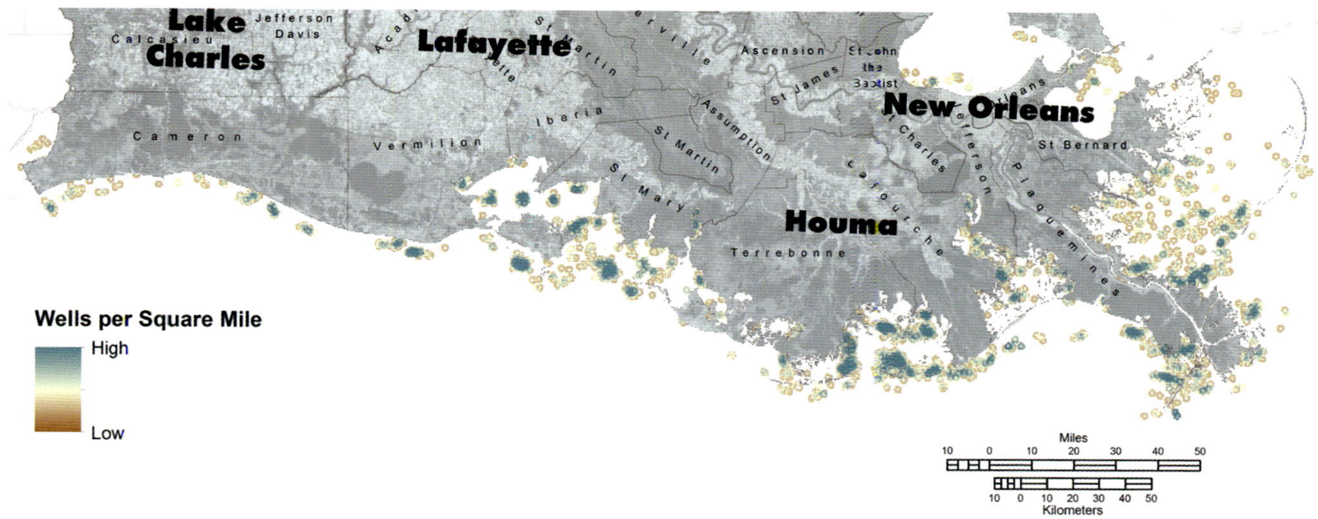

Fig. 54. Density of Nearshore Oil Wells (2010)

Data Source: Louisiana Department of Natural Resources

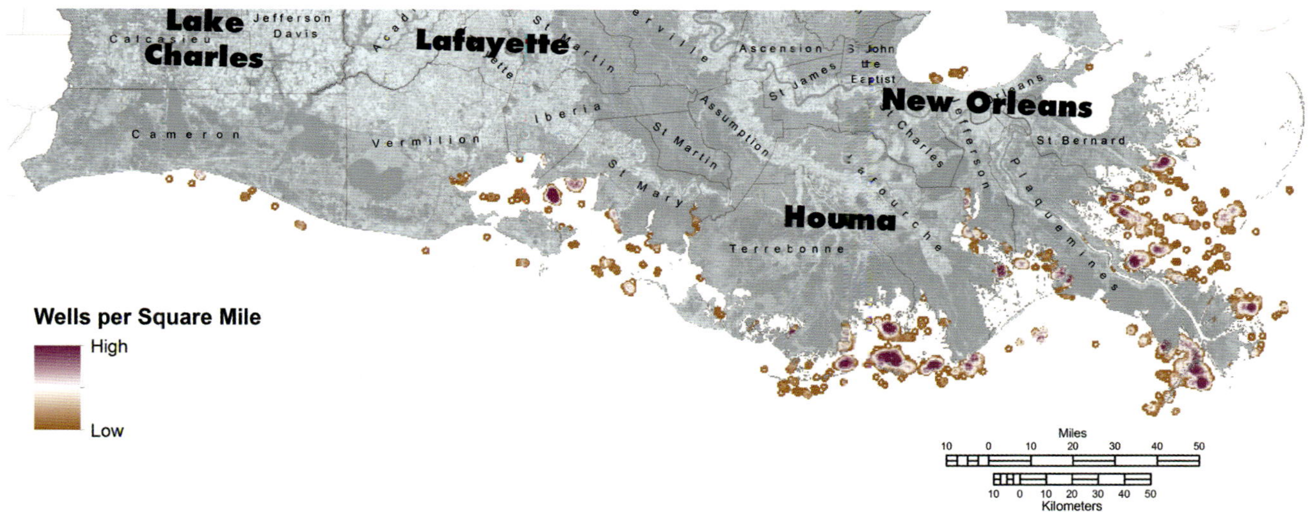

This onshore production largely peaked by the 1970s, although new technologies would allow for additional development up to the present. This new development is occurring both onshore, in the form of hydraulic fracturing to extract shale oil in northwest Louisiana, and offshore, with the move into ever-deeper waters (figs. 53-54). The first well drilled over open water was in Caddo Lake in 1910. Later advances allowed for offshore drilling in the Gulf of Mexico. In 1947, just south of Morgan City, the first offshore well was drilled out of sight of land. Discovery of the West Cameron Field in 1954 established Louisiana's offshore oil industry

Fig. 55. Density of OCS Oil and Gas Platforms (2010)
Data Source: U.S. Bureau of Ocean Energy Management

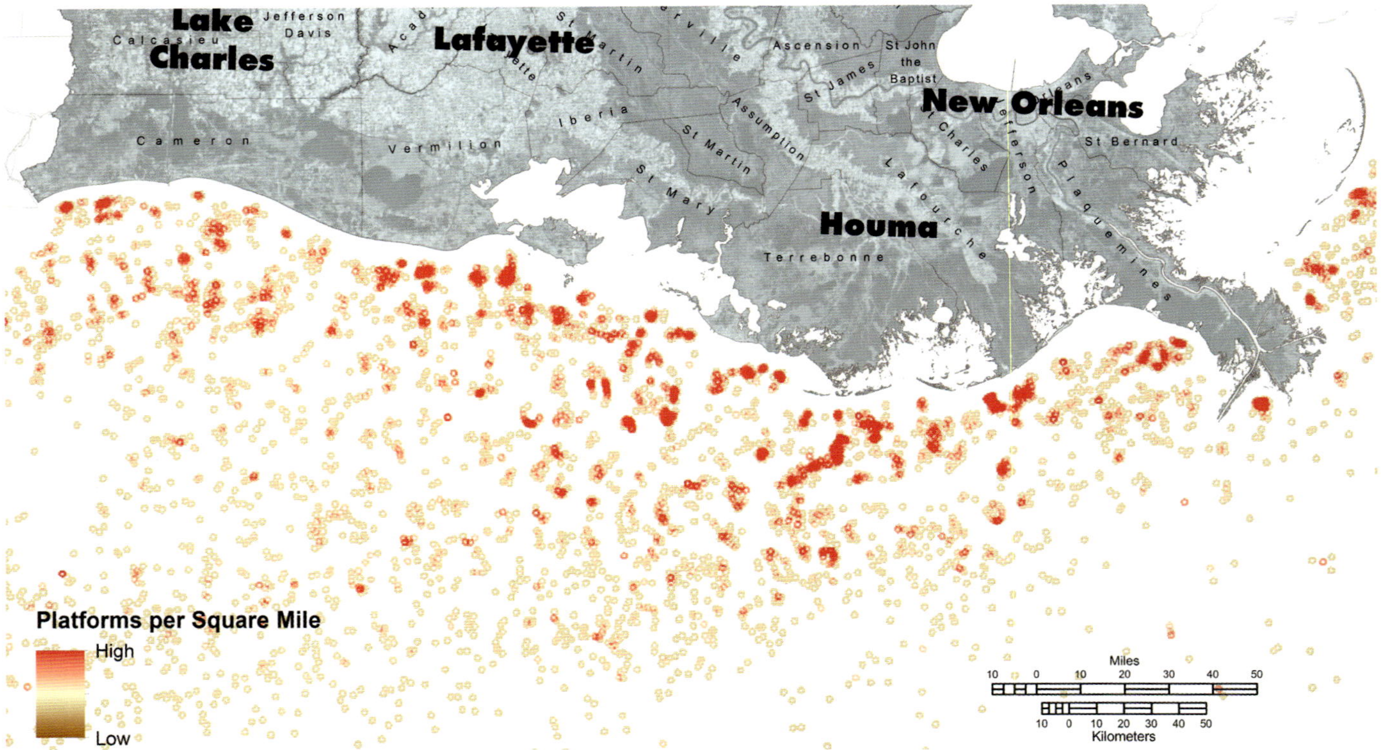

The final frontier of oil and gas exploration would be on the OCS and in the deep water of the Gulf of Mexico. Advances in seismic technology as well as the development of non-fixed oil platforms in the 1980s made it economically feasible to drill far out in the Gulf (fig. 55). As a consequence, a new offshore-oriented industry began to grow in coastal Louisiana.

MALADAPTIVE PRACTICES IN THE COASTAL ZONE

While communities have found effective ways to adapt to changing conditions in Louisiana's coastal zone, other adaptations have proven to be less successful. In certain cases, communities have adapted in ways ill-suited to life in the coastal zone. Some maladaptive practices have been driven by broader social and economic forces while others have been driven by outside social policies.

Fig. 56. Minority Populations Residing in SFHAs in Lake Charles, Louisiana

Because most of Lake Charles is outside of the floodplain, minority residents generally do not reside in locations where they will be disproportionately affected by heightened flood risk.

Data Sources: FEMA Flood Insurance Rate Maps; U.S. Census Bureau

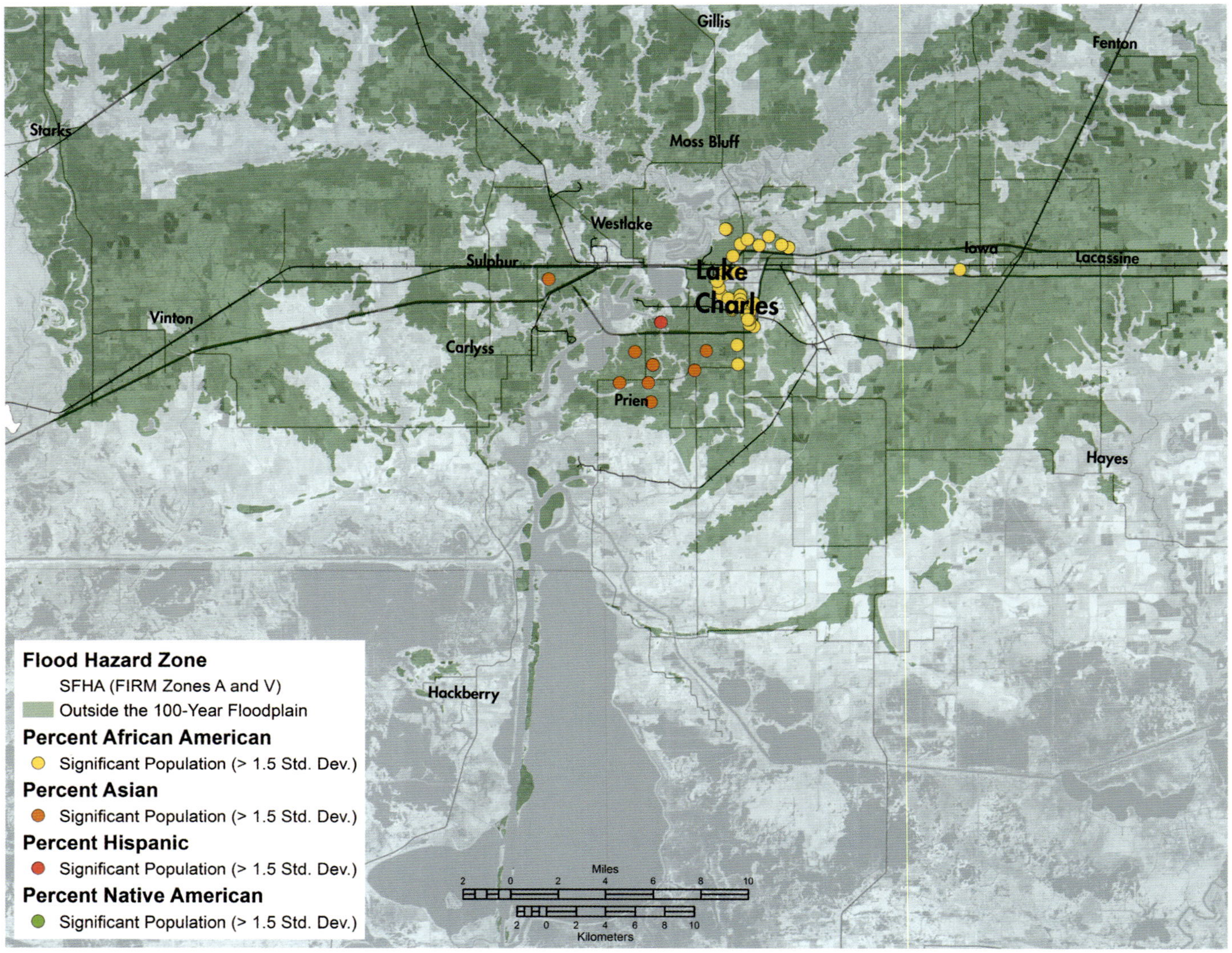

Chief among the outside social policies is the National Flood Insurance Program (NFIP), a program developed in 1968, in response to the impacts of Hurricane Betsy in 1965, to provide subsidized federal flood insurance to individuals residing in designated flood hazard areas (figs. 56–59). In 1972, following Hurricane Agnes, the purchase of flood insurance became mandatory for all mortgaged properties located in FEMA-designated Special Flood Hazard Areas (SFHAs). While the NFIP was originally intended to reduce floodplain development and risk, it essentially allowed property owners to develop land deemed too risky for private insurance to cover. Despite the original intentions of the program, a vast majority of Louisianans residing in the coastal zone have federally subsidized flood insurance. In the years following the storms of 2005, including Hurricanes Katrina, Rita, and Wilma, federal officials raised questions about the solvency of the program and discussed efforts to raise flood insurance premiums. Should the government raise premiums by any significant amount, the impacts on coastal communities could be immeasurable, particularly for those low-income and minority residents already living on the edge of poverty.

Fig. 57. Minority Populations Residing in SFHAs in Lafayette, Louisiana

In Lafayette and surrounding communities, most of the population resides outside of the 100-year floodplain. However, development in low-lying areas places groups that reside in these areas, such as the Hispanic population, at enhanced risk, particularly if these residents are new immigrants or use English as a second language.

Data Sources: FEMA Flood Insurance Rate Maps; U.S. Census Bureau

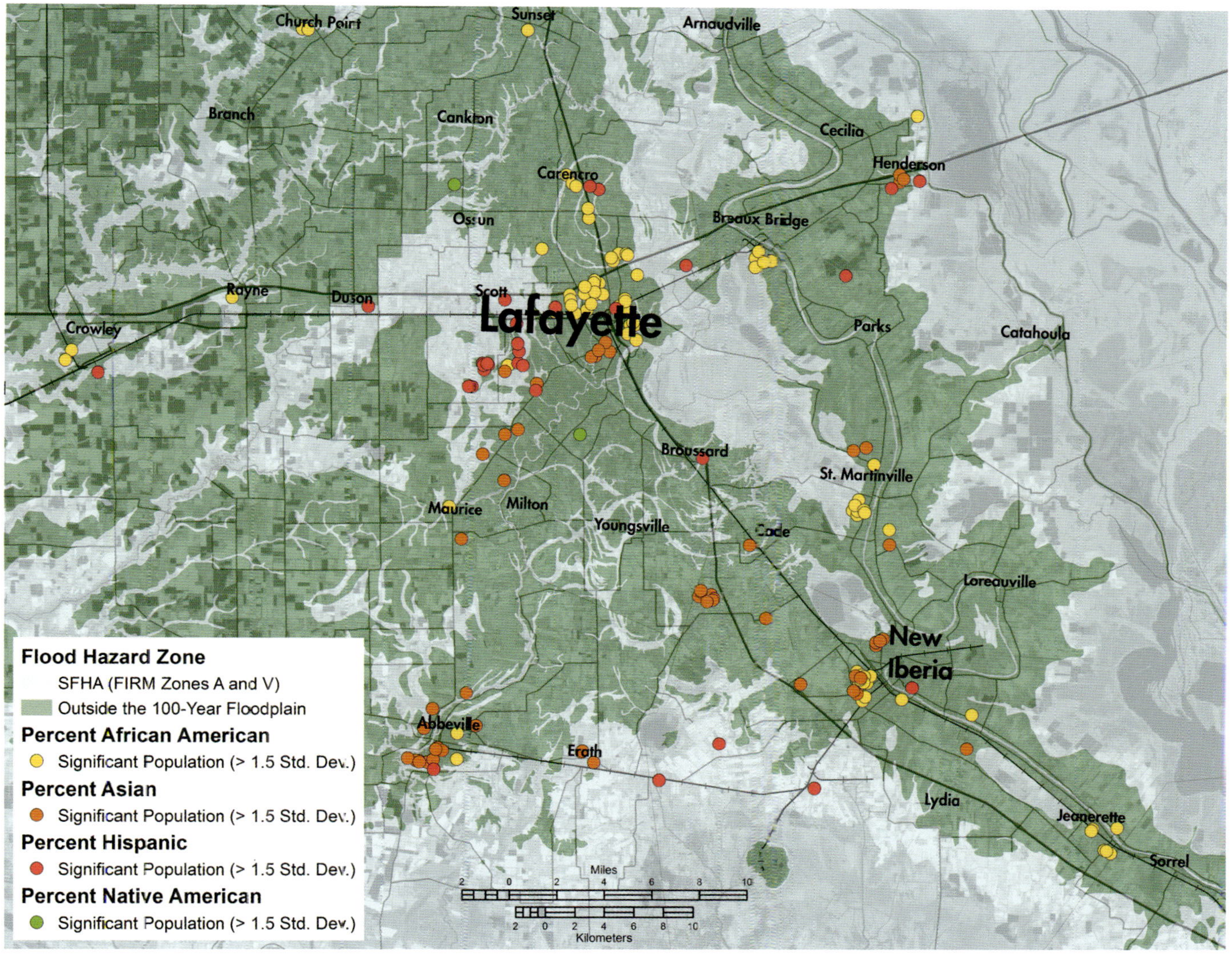

Fig. 58. Minority Populations Residing in SFHAs in Houma, Louisiana

Terrebonne and Lafourche parishes represent the southern fringe of habitable land in southeast Louisiana. This area is also home to a large portion of the state's Native American population, who reside mainly in communities such as Dulac, Chauvin, and Golden Meadow atop ground that slowly drops in elevation as the land gradually slopes toward the Gulf of Mexico. Although some of these areas are well protected by fully enclosed levee polders, as with the Lafourche Parish communities from Lockport to Golden Meadow, the region's low elevation means that these areas are still at increased flood risk.

Data Sources: FEMA Flood Insurance Rate Maps; U.S. Census Bureau

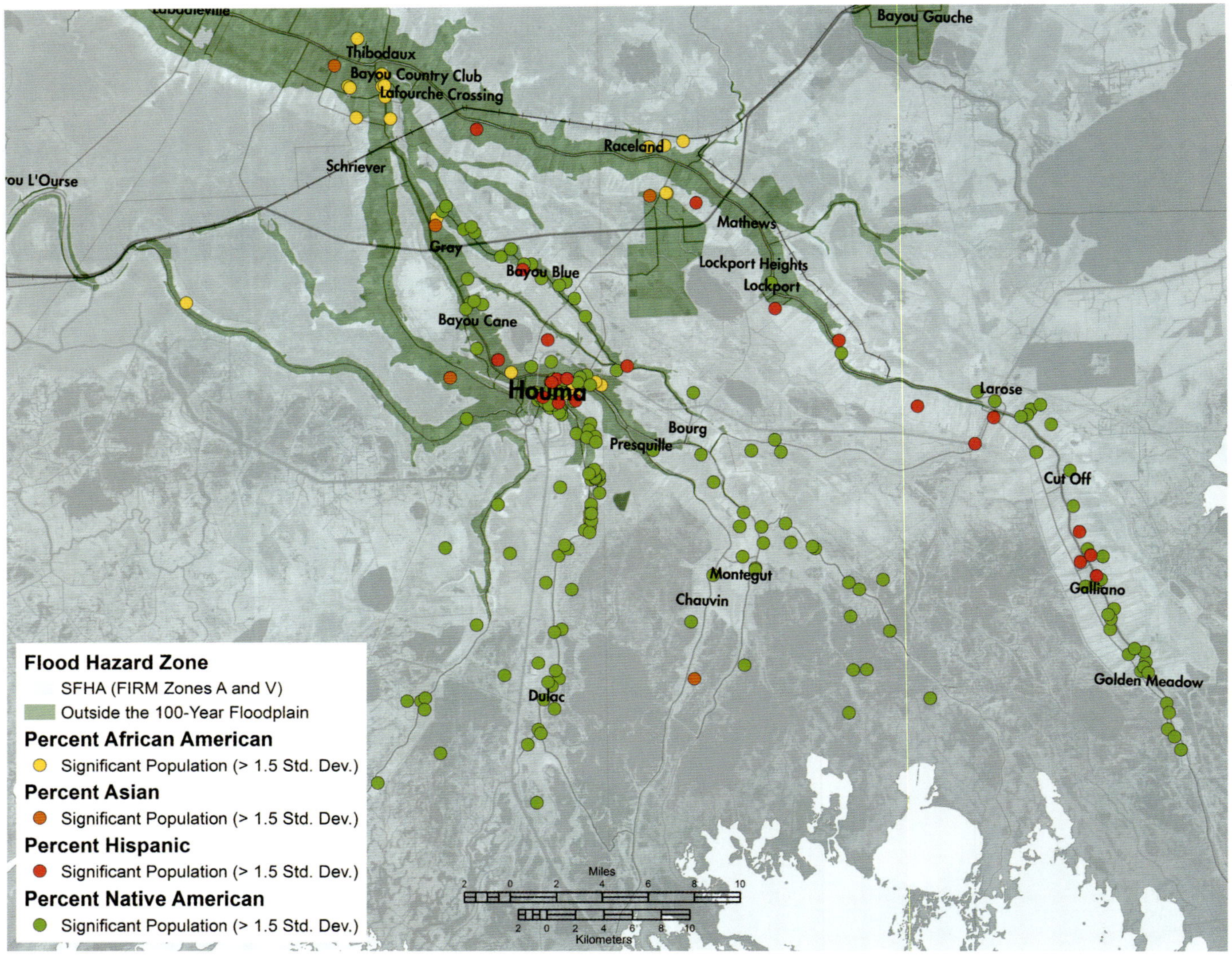

Fig. 59. Minority Populations Residing in SFHAs in New Orleans, Louisiana

Special mention should be made of the New Orleans metropolitan area. New Orleans is the largest, most densely populated city in the state, and one of the largest on the Gulf Coast. The city also has one of the most advanced levee protection systems in the country. Yet most of New Orleans's residents reside in SFHAs.

Data Sources: FEMA Flood Insurance Rate Maps; U.S. Census Bureau

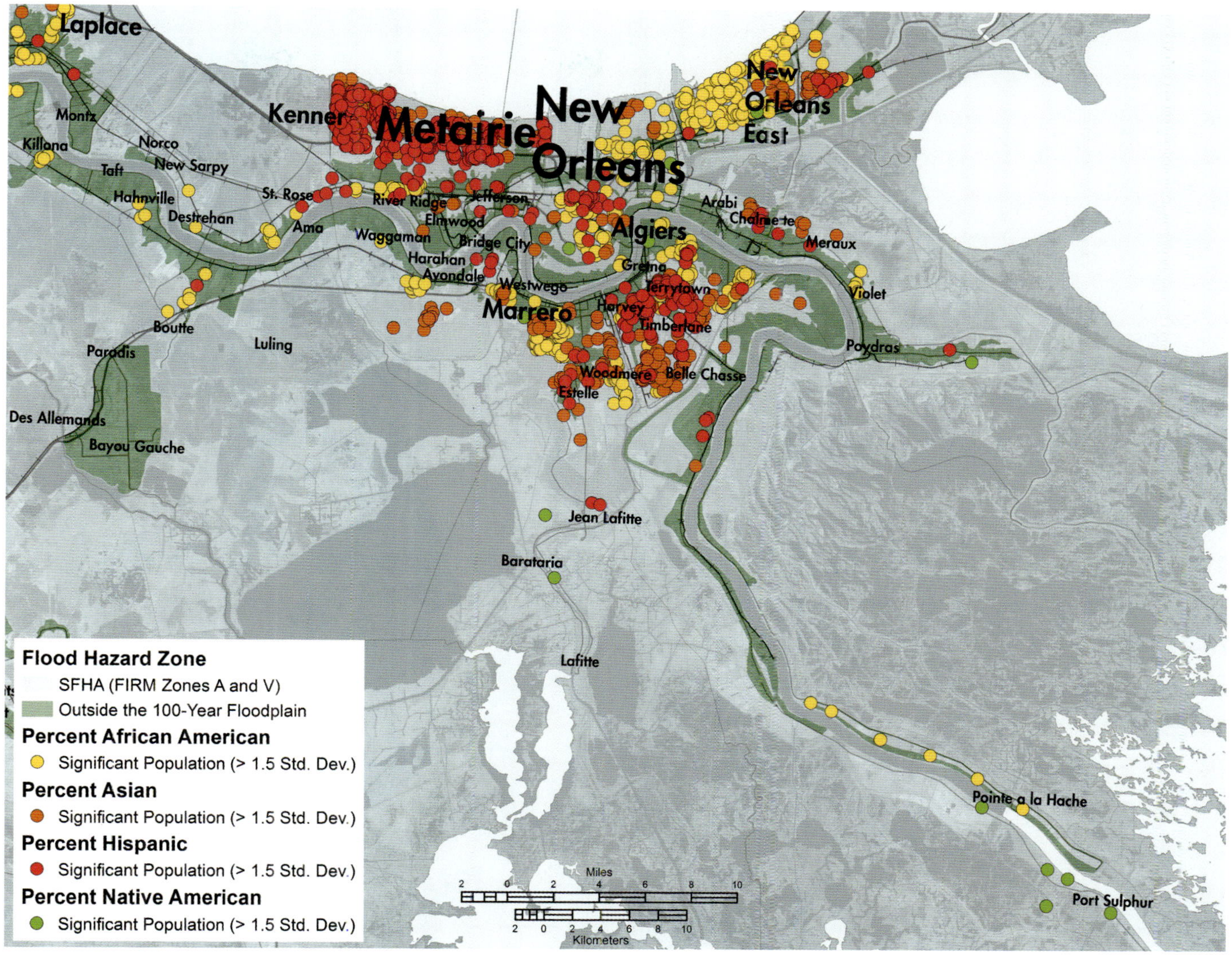

Fig. 60. Percentage of Houses Built on Concrete Slabs (1960)

Data Source: U.S. Census Bureau

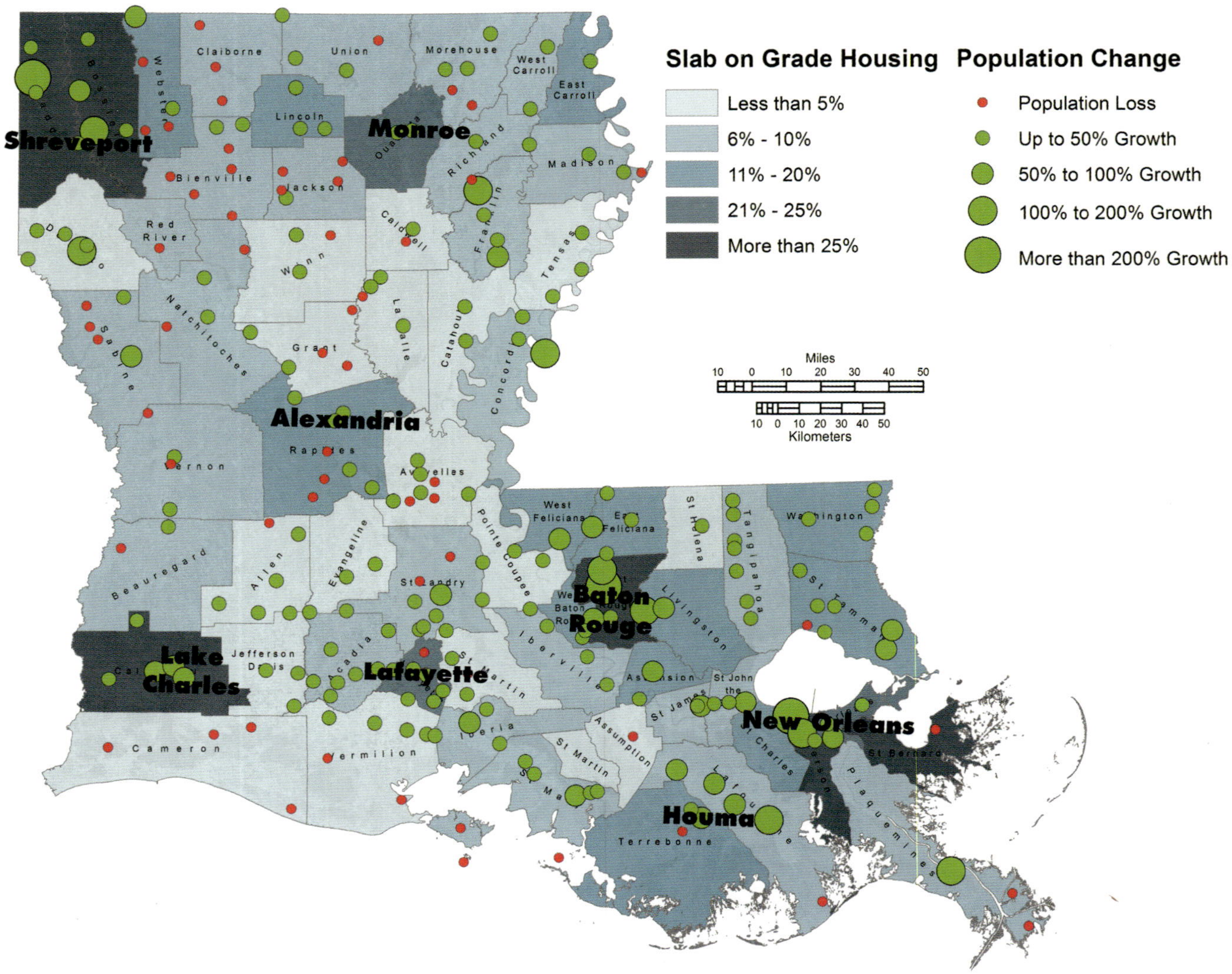

Other changes to the social and built landscape have had similar deleterious impacts on the levels of risk faced by coastal residents. Historically, elevated homes provided a degree of nonstructural protection as most vernacular house types in older areas of the state were built on brick or concrete-block pier foundations. The post-World War II housing boom reversed this trend; much of the new development, which tended to occur on low ground far from the natural levees, consisted of modern construction on slab-on-grade foundations (fig. 60). This maladaptive construction combined with settlement in high-risk areas has resulted in a state of heightened vulnerability for many residents of coastal Louisiana.

POPULATION MIGRATION

The adaptive measure often viewed as one of the last resorts for coastal residents is population movement and relocation. Population movement has long been a viable strategy for residents of coastal and noncoastal areas alike. Though not traditionally thought of as relocation, populations have often moved to other communities, states, or even countries in search of better opportunities. Louisiana is no different. Throughout the latter half of the twentieth century, populations across the state moved into and out of various regions for any number of reasons.

Fig. 61. Population Change in Louisiana Communities by Decade (1950–2010)

6 maps

Data Source: U.S. Census Bureau

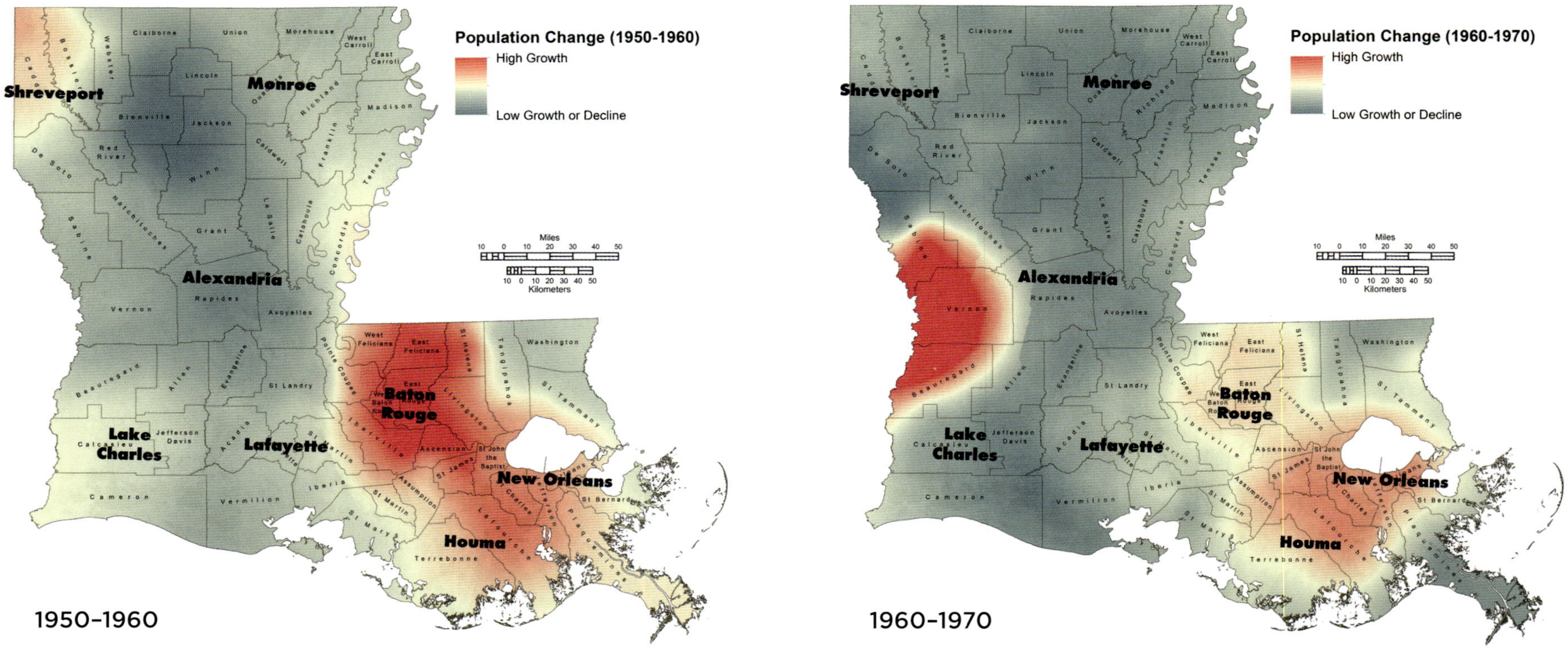

Some of the reasons for population shifts (fig. 61) are related to nationwide societal trends, such as the urbanization trend of the 1950s and the suburbanization of the 1960s and 1970s. In coastal Louisiana, this suburbanization focused largely on the area around Lake Pontchartrain, including Jefferson and St. Bernard parishes. Other population trends were driven by specific events, such as the growth of population in the Fort Polk area during the Vietnam War era or the growth of Lafayette Parish at the start of the offshore oil boom of the 1990s. The impact of natural disasters cannot be overstated. Population changes occurring between 2000 and 2010 revolved around the impacts of Hurricanes Katrina and Rita. Highly significant population losses

Fig. 61. Population Change in Louisiana Communities by Decade (1950–2010)
Data Source: U.S. Census Bureau

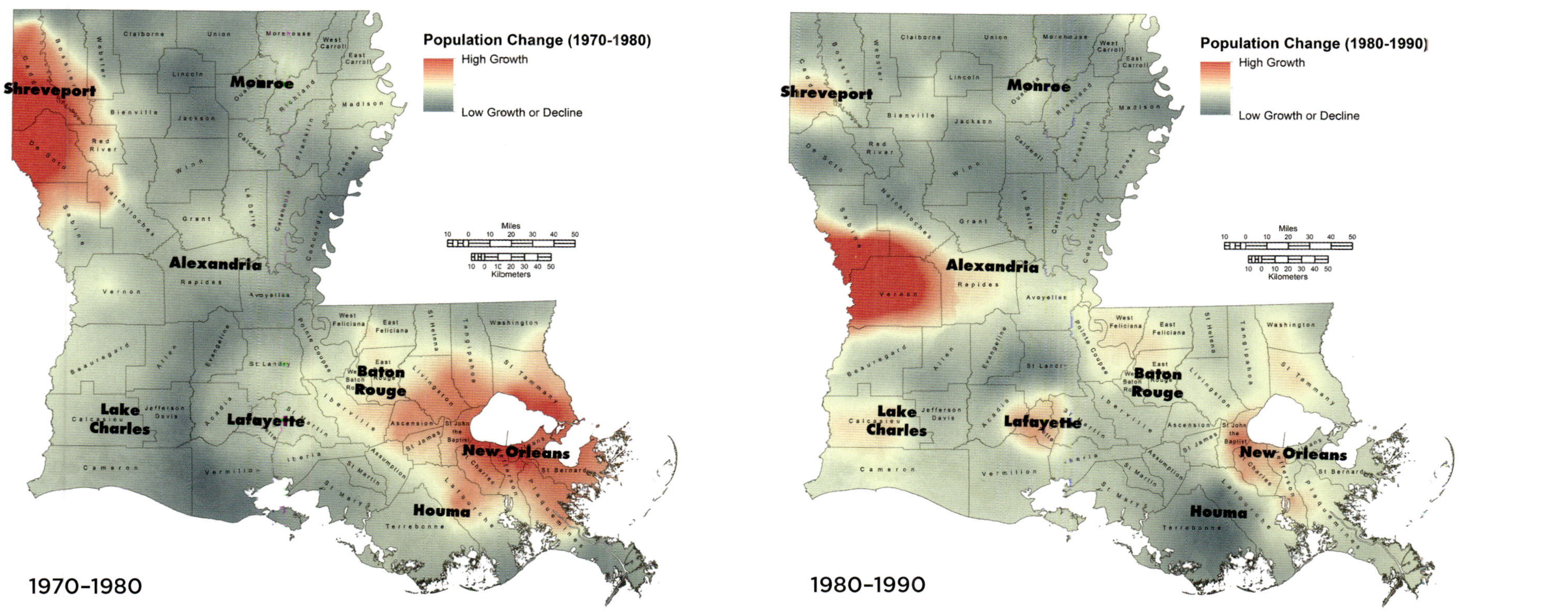

are evident in the New Orleans metro area, including all of the immediate neighboring parishes, as well as in Cameron Parish in southwest Louisiana, which was heavily impacted by Hurricane Rita. Population growth over this time occurred primarily in communities on the north shore of Lake Pontchartrain and in suburban Baton Rouge. In fact, experiencing the greatest growth during this time period were the suburban parishes surrounding Baton Rouge, where the impacts of suburbanization combined with a movement away from New Orleans and the broader coastal zone (fig. 62).

Fig. 61. Population Change in Louisiana Communities by Decade (1950–2010)

Data Source: U.S. Census Bureau

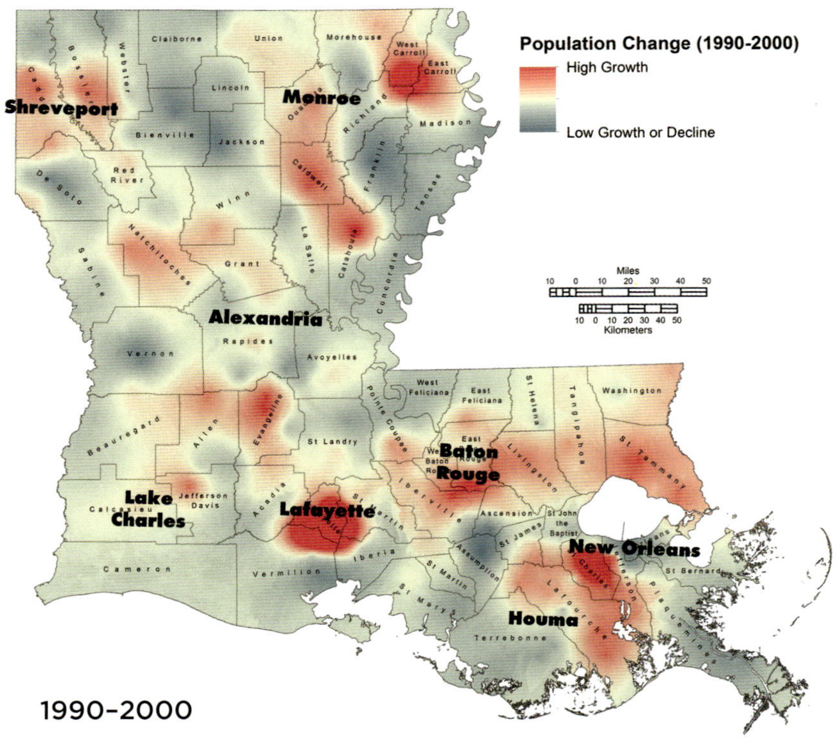

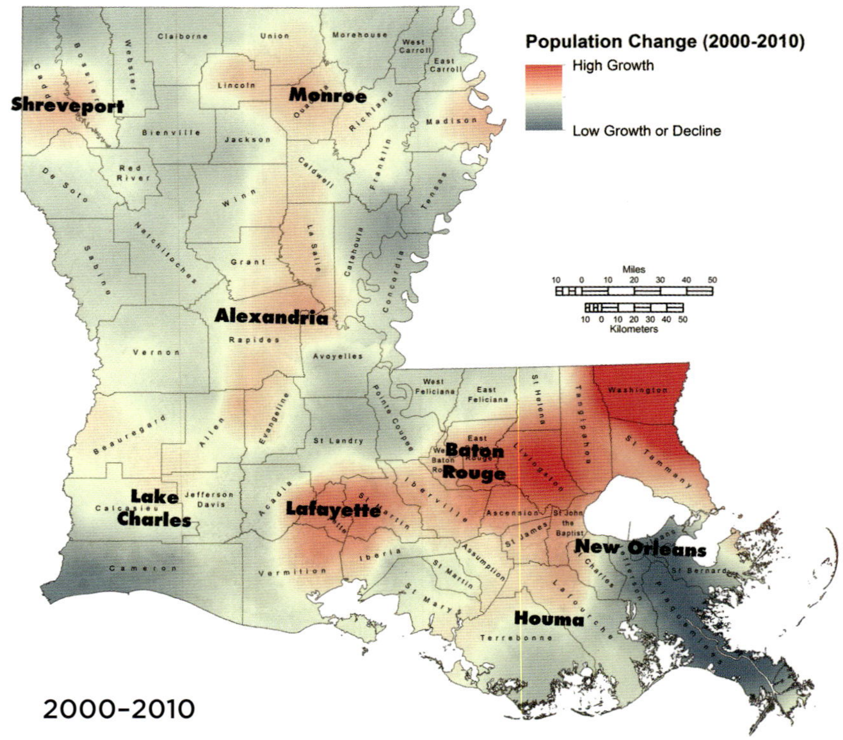

Fig. 62. Total Population of Communities in Coastal Shoreline and Coastal Watershed Parishes (1950–2010)

7 maps

Data Source: U.S. Census Bureau

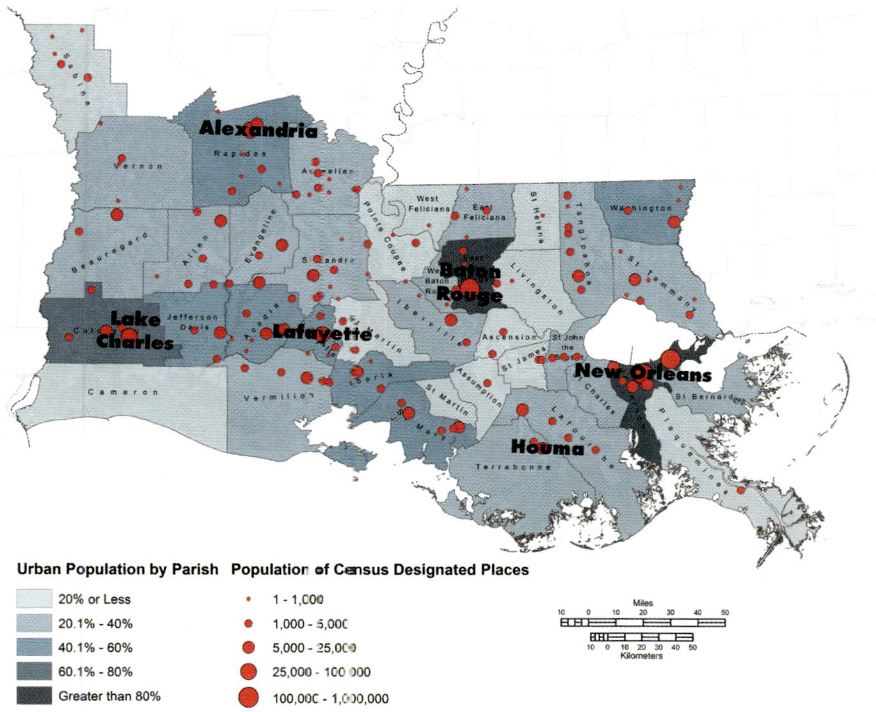

1950

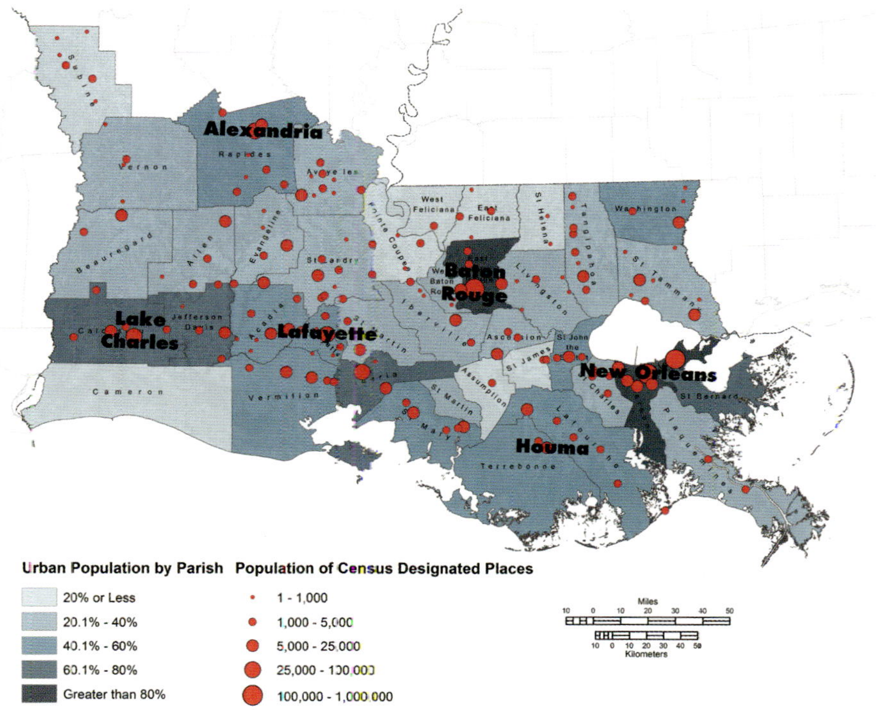

1960

Despite the many population changes that occurred in the latter half of the twentieth century and in the first decade of the twenty-first, total population patterns remained fairly constant with the majority of the state's coastal population residing in the most urbanized parishes. Most of the new growth occurred along the I-10 corridor between New Orleans and Baton Rouge and the Highway 90 corridor between Houma and Lafayette.

Fig. 62. Total Population of Communities in Coastal Shoreline and Coastal Watershed Parishes (1950–2010)

Data Source: U.S. Census Bureau

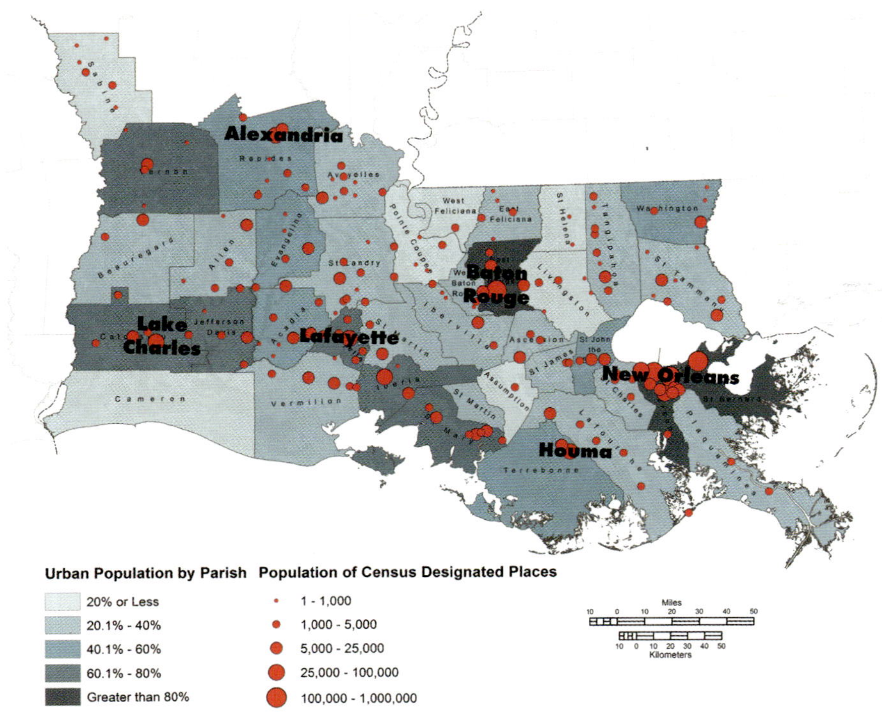

1970

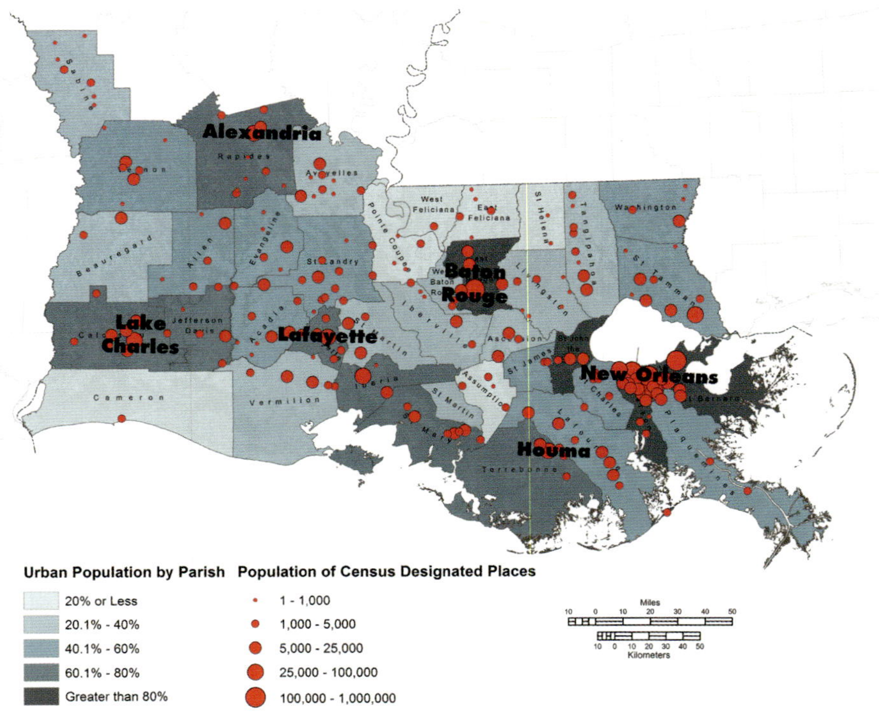

1980

Fig. 62. Total Population of Communities in Coastal Shoreline and Coastal Watershed Parishes (1950–2010)

Data Source: U.S. Census Bureau

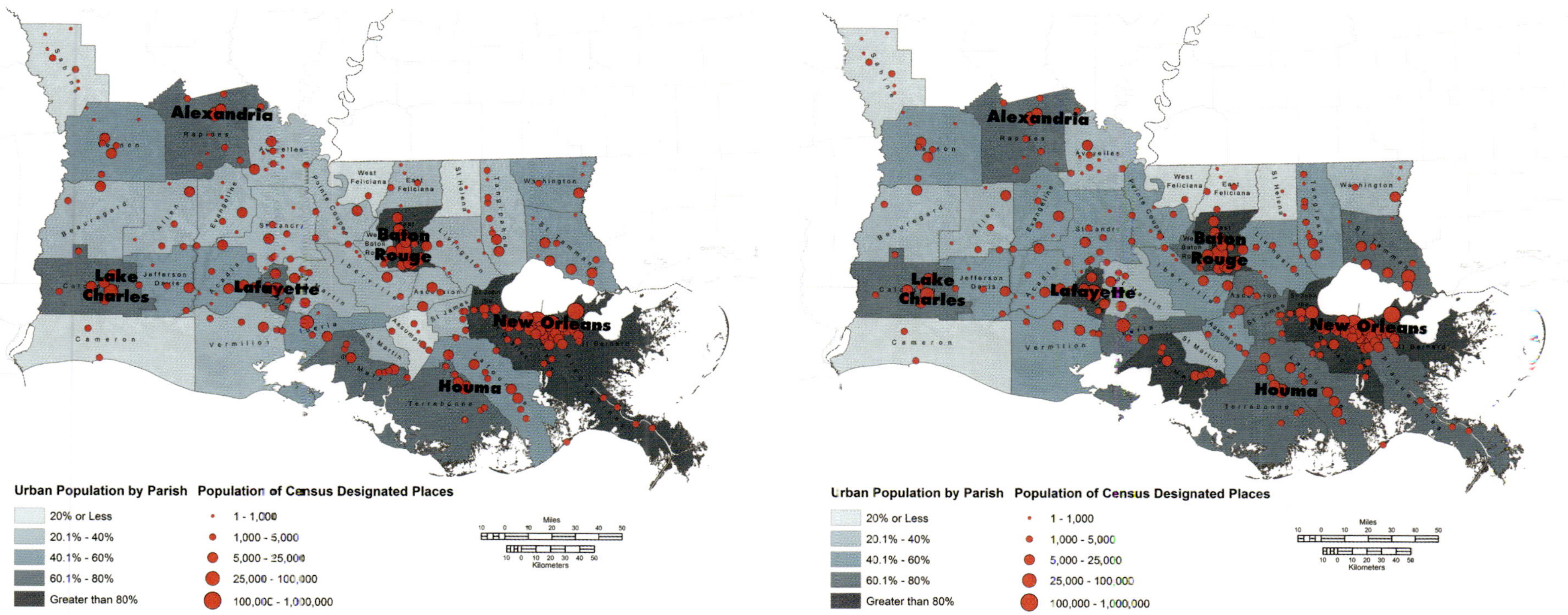

1990

2000

Fig. 62. Total Population of Communities in Coastal Shoreline and Coastal Watershed Parishes (1950–2010)

Data Source: U.S. Census Bureau

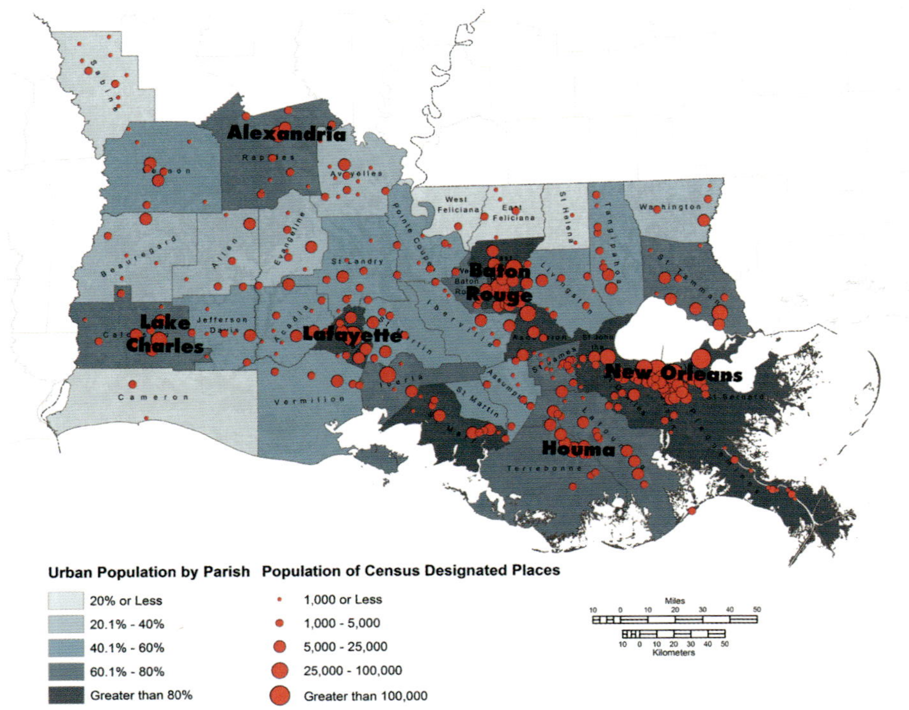

2010

Fig. 63. Population Moving from Another State during the Previous Decade

6 maps

Data Source: U.S. Census Bureau. (*Note:* The American Community Survey [ACS] replaced the Decennial Census in 2010. The ACS uses a 5-year rolling average, and the Previous Residence question is based on the previous year's residence rather than five years earlier.)

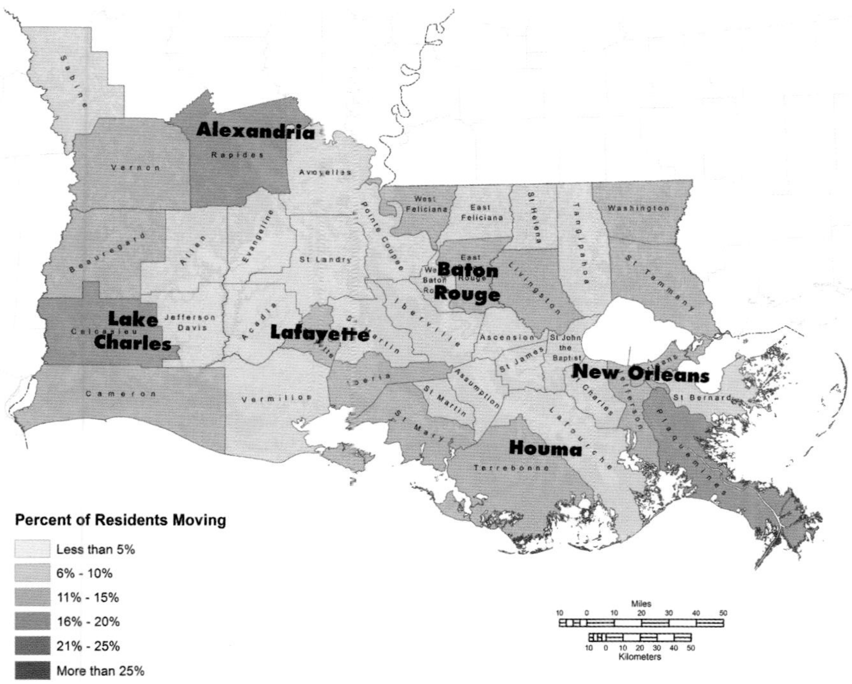

1955–1960

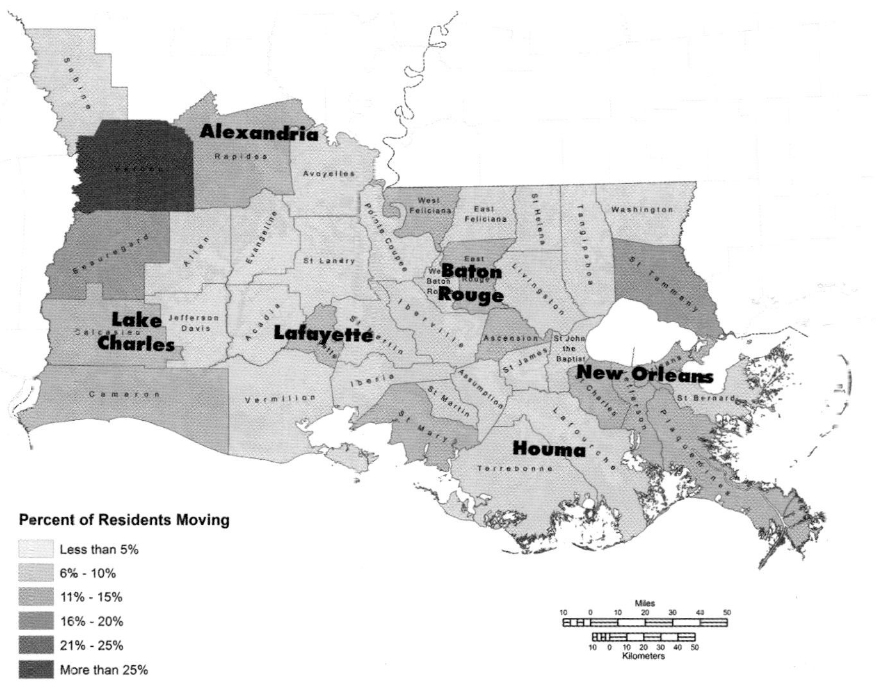

1965–1970

Very little population growth in Louisiana appears to be the result of migration from out of state (fig. 63). The exception to this is the region surrounding Fort Polk, a military base in Vernon Parish, at the northwestern edge of Louisiana's coastal watershed area. From the 1960s onward, Vernon Parish has drawn a quarter of new residents from out of state. A spillover effect of this out-of-state population growth is also seen in Vernon's neighboring parishes.

Fig. 63. Population Moving from Another State during the Previous Decade

Data Source: U.S. Census Bureau. (*Note:* The American Community Survey [ACS] replaced the Decennial Census in 2010. The ACS uses a 5-year rolling average, and the Previous Residence question is based on the previous year's residence rather than five years earlier.)

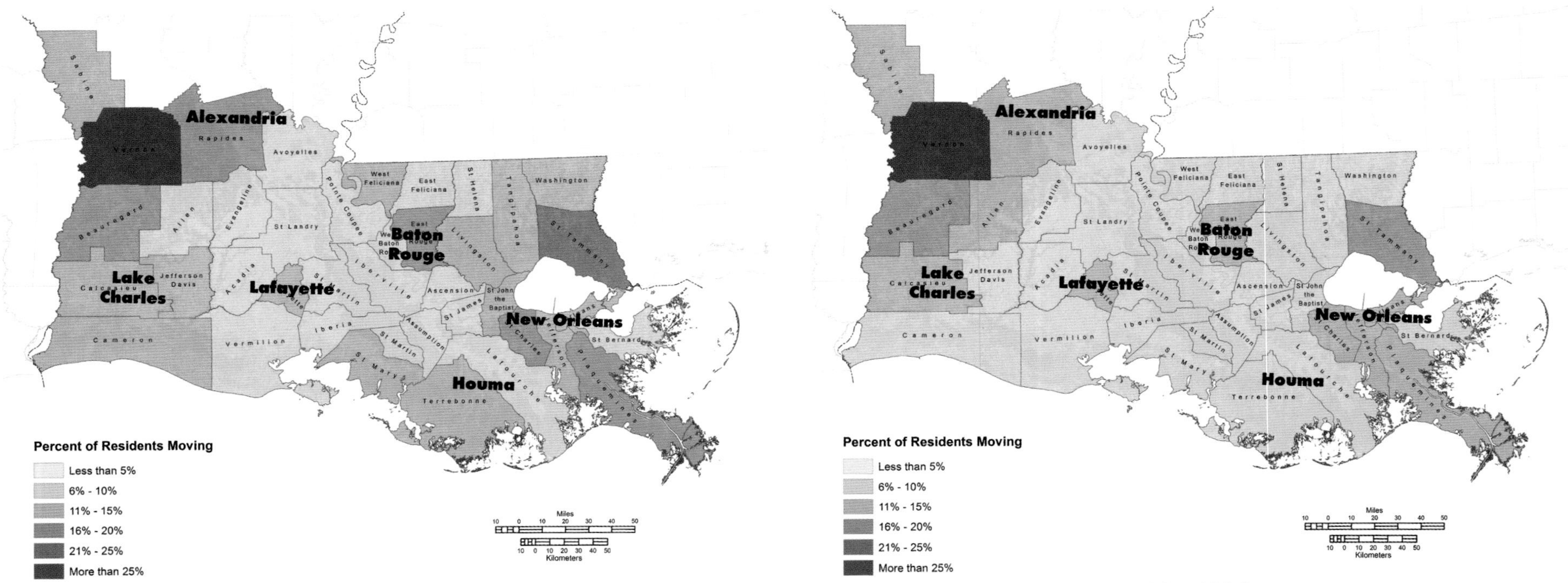

1975–1980

1985–1990

Fig. 63. Population Moving from Another State during the Previous Decade

Data Source: U.S. Census Bureau. (*Note:* The American Community Survey [ACS] replaced the Decennial Census in 2010. The ACS uses a 5-year rolling average, and the Previous Residence question is based on the previous year's residence rather than five years earlier.)

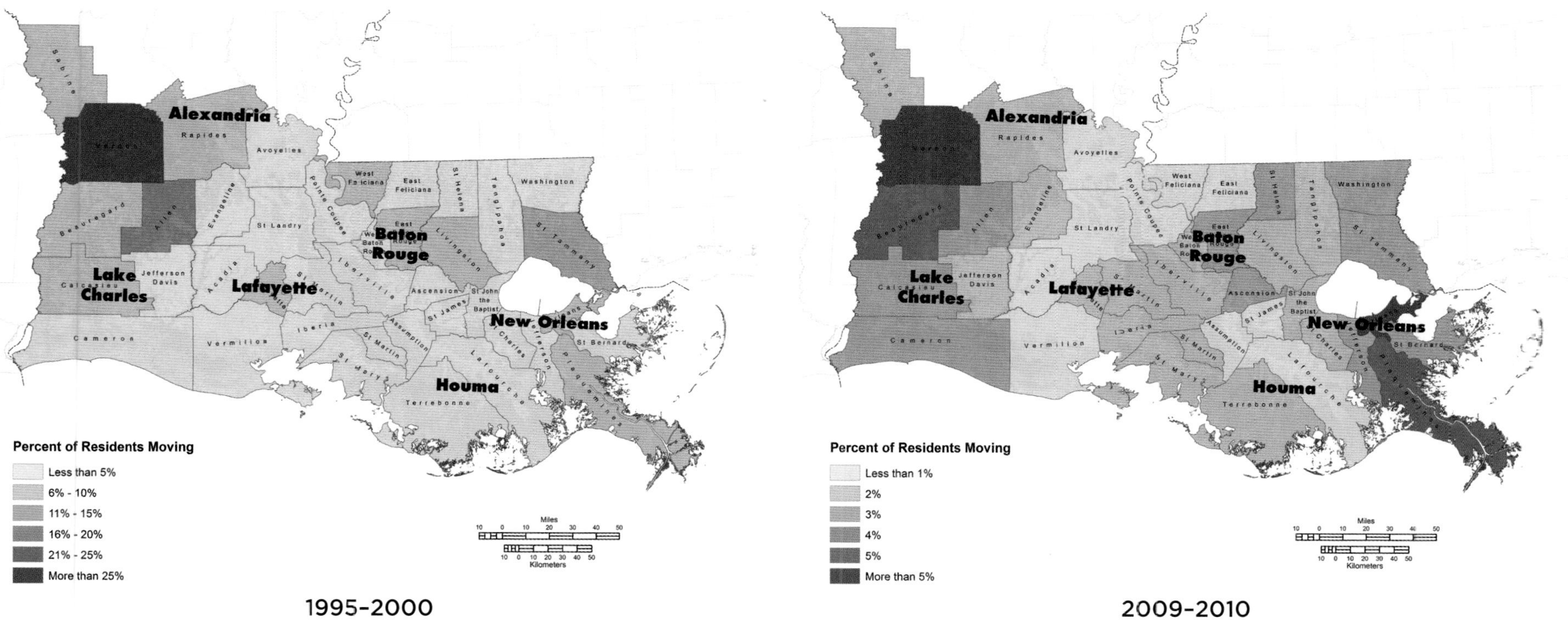

Fig. 64. Population Moving from Another Parish during the Previous Decade

6 maps

Data Source: U.S. Census Bureau. (*See note to figure 63.*)

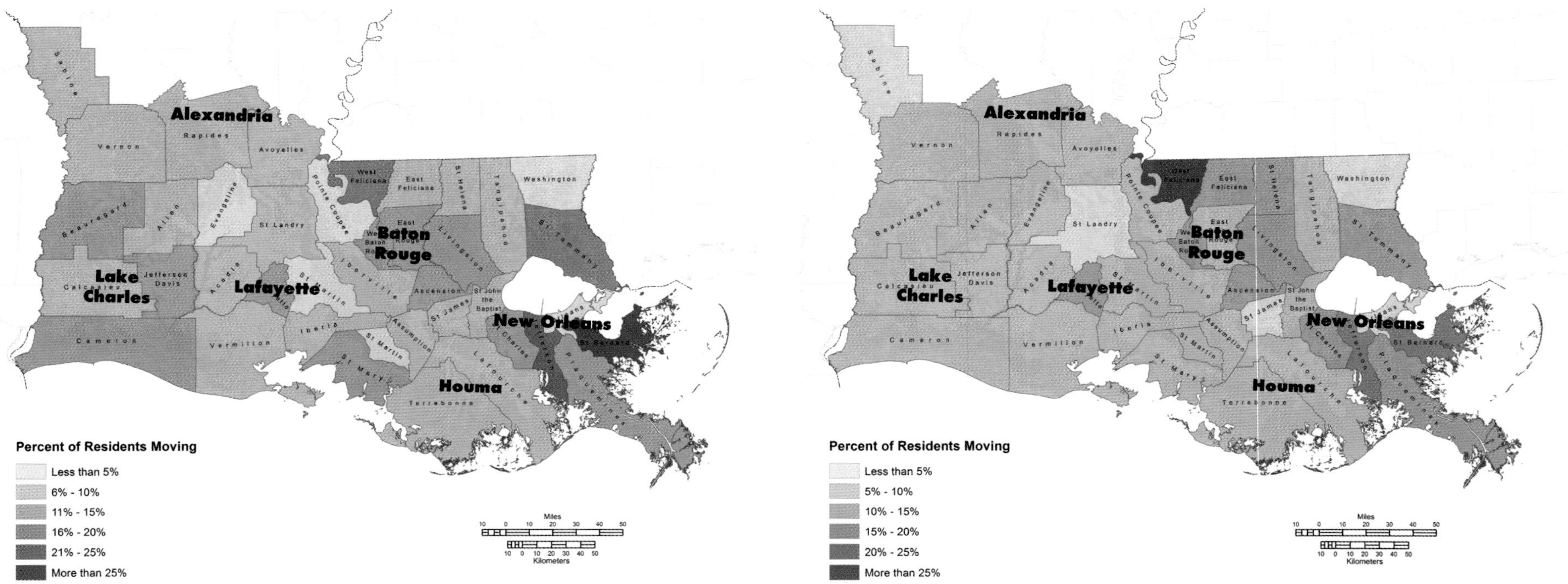

1955–1960

1965–1970

Movement from one parish to another in Louisiana is much more common, although there are some regional patterns that stand out (fig. 64). First, during the late 1950s, the dispersion of population from New Orleans into the neighboring suburban parishes of Jefferson and St. Bernard began the process that would continue for the remainder of the century, the development of the circum-Pontchartrain region. This process would continue to some degree up until the present. This movement from one parish to another was the impetus for the growth of the Northshore parishes and suburban Baton Rouge, especially during the 1970s. It is particularly notable that much of this population dispersion occurs along the Interstate 10 and Interstate 12 corridors. By the late 1990s, population began to expand along the Interstate 10 corridor toward Lafayette Parish, in the central part of the coastal area. In general, this process suggests that the development of interstate highway networks have begun to transform much of the coastal parishes into a more interconnected region.

Fig. 64. Population Moving from Another Parish during the Previous Decade

Data Source: U.S. Census Bureau. (*See note to figure 63.*)

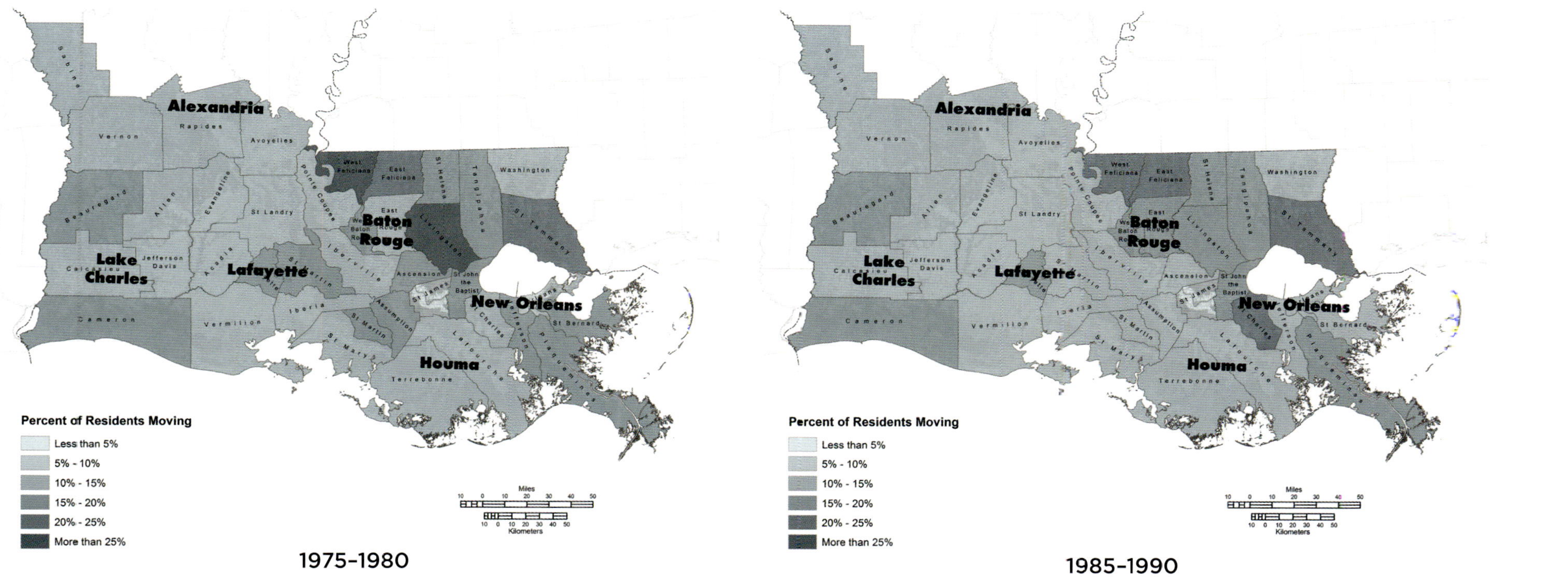

Fig. 64. Population Moving from Another Parish during the Previous Decade

Data Source: U.S. Census Bureau. (*See note to figure 63.*)

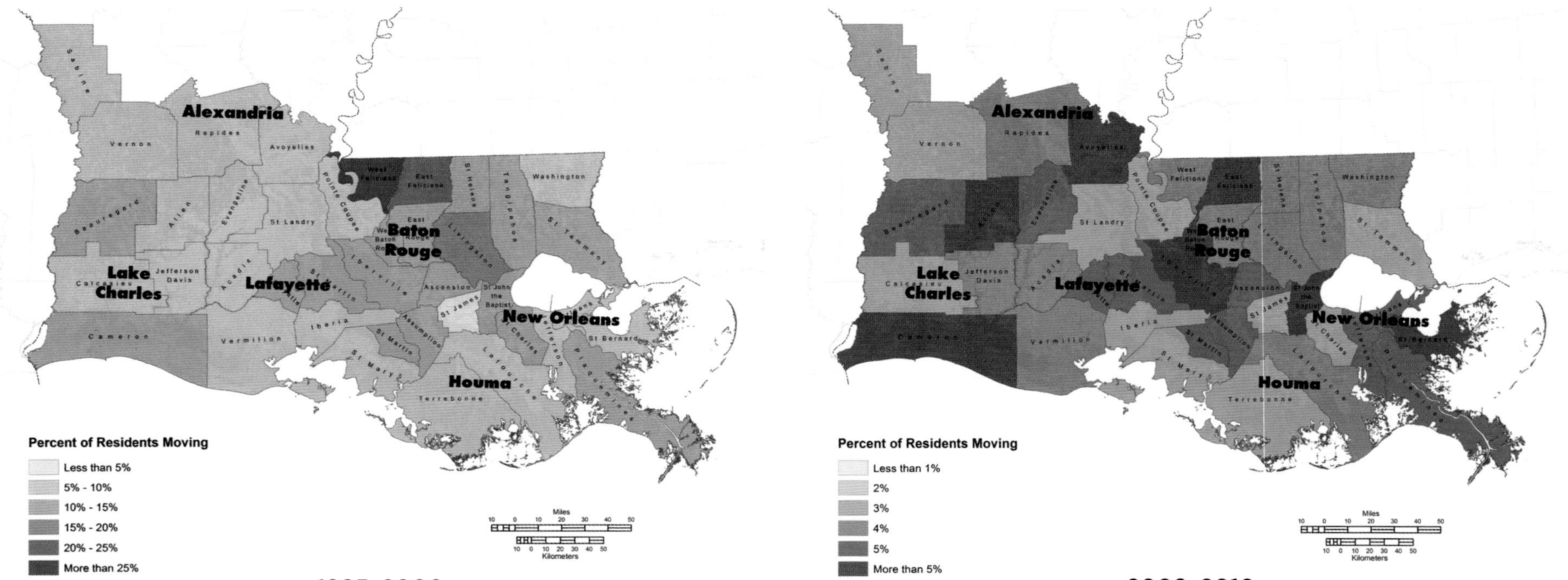

1995–2000

2009–2010

Fig. 65. Population Residing in the Same House during the Previous Decade

6 maps

Data Source: U.S. Census Bureau. (*See note to figure 63.*)

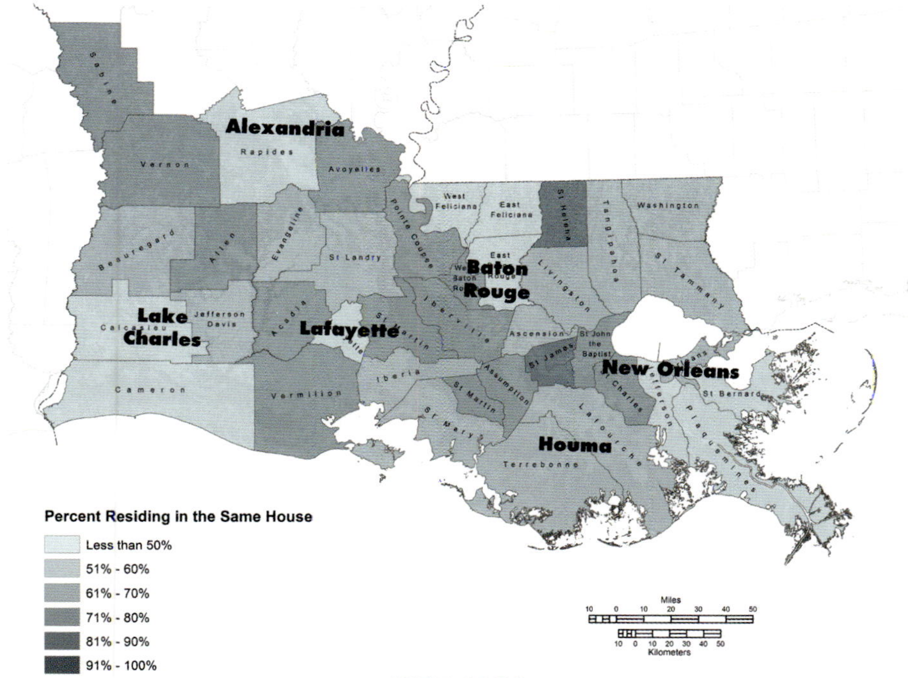

1955–1960

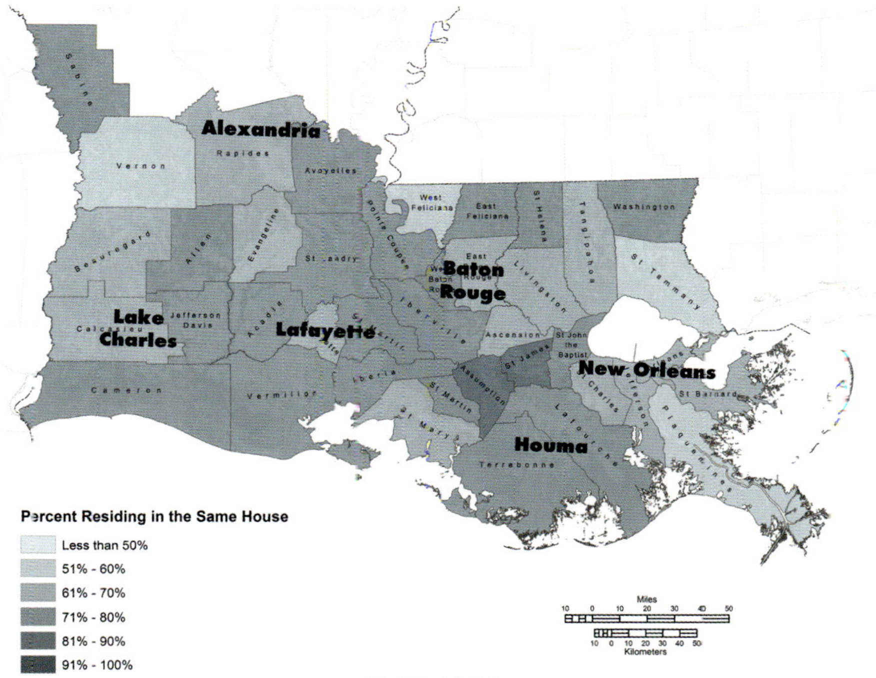

1965–1970

As with many population variables, any number of factors can impact why populations remain in or leave particular areas and why they are drawn to others. Population stability, represented by the number of residents living in the same home from decade to decade (fig. 65), is often used as an indicator of place attachment, although it may also be an indicator of low economic growth. In coastal Louisiana, the rural-agricultural core of the region, stretching across the western portion of the state between Baton Rouge and Lake Charles, contains the largest proportion of residents who remain in the same home from one census period to the next, with the exception of the more urbanized Lafayette Parish.

Fig. 65. Population Residing in the Same House during the Previous Decade

Data Source: U.S. Census Bureau. (*See note to figure 63.*)

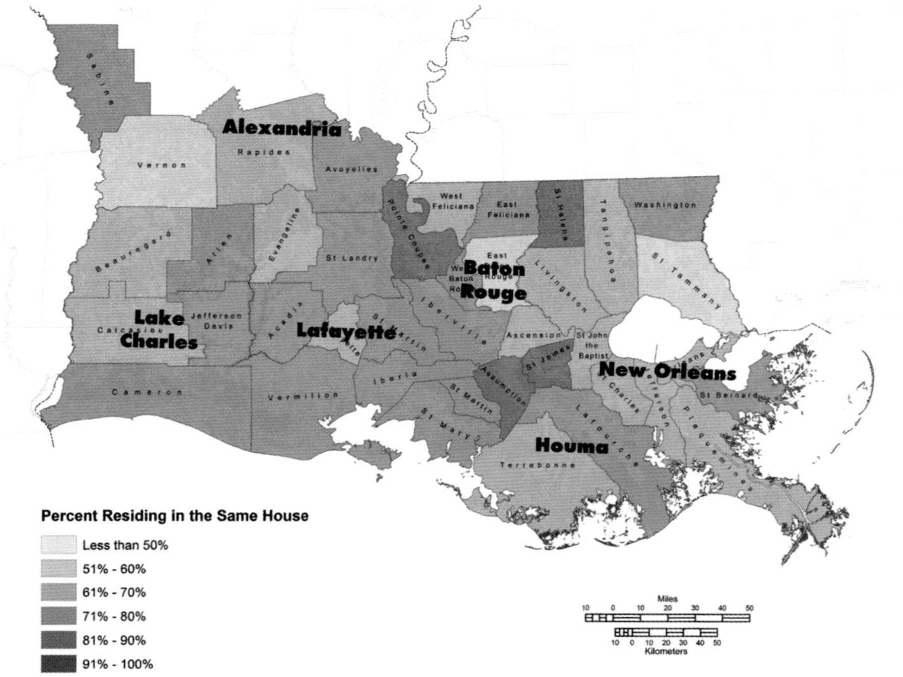

1975–1980

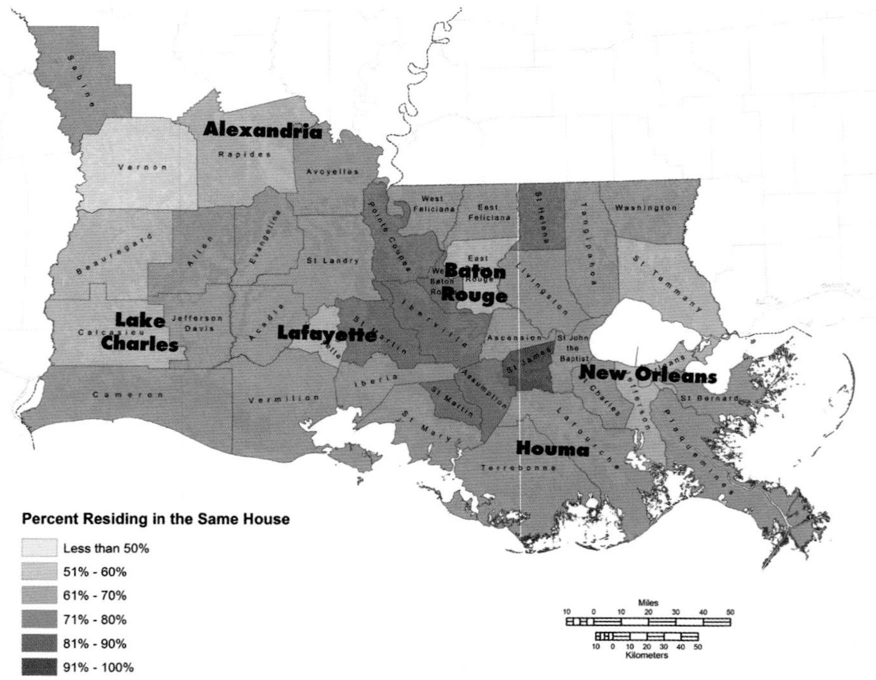

1985–1990

Fig. 65. Population Residing in the Same House during the Previous Decade

Data Source: U.S. Census Bureau. (*See note to figure 63.*)

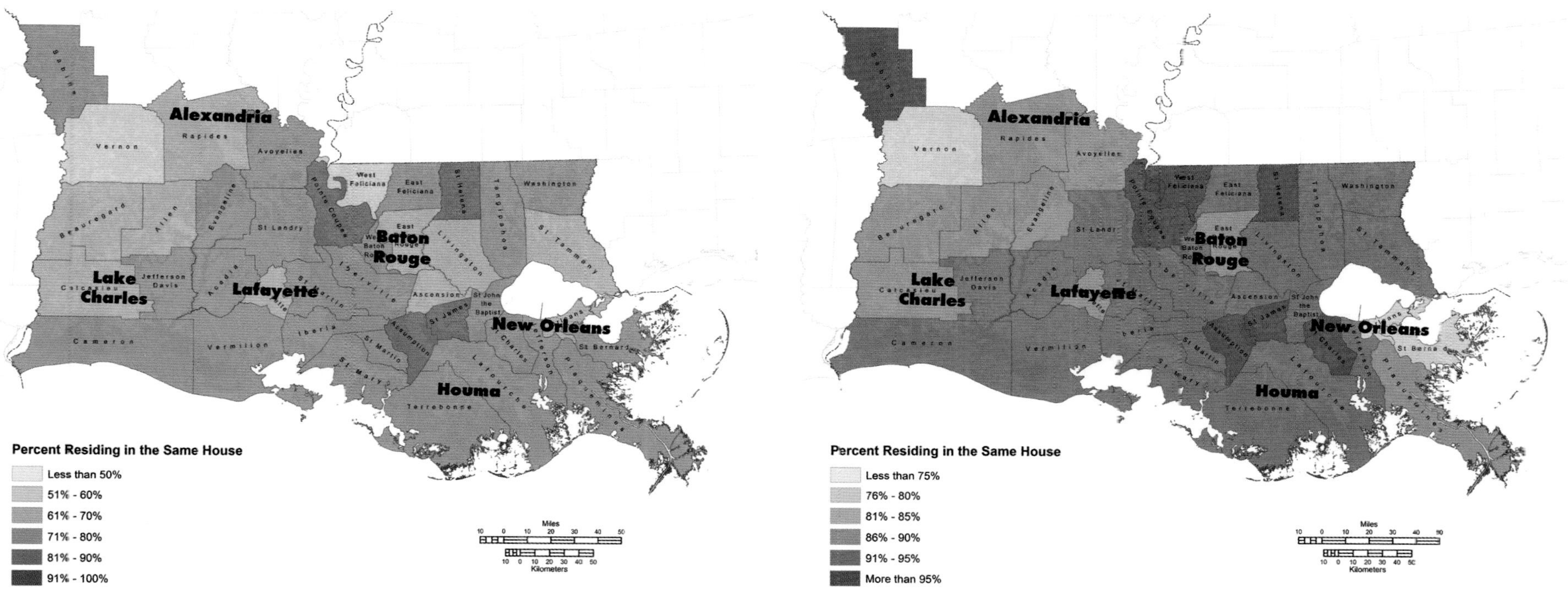

1995–2000

2009–2010

Fig. 66. Vacant Housing in Coastal Shoreline and Coastal Watershed Parishes

7 maps

Data Source: U.S. Census Bureau

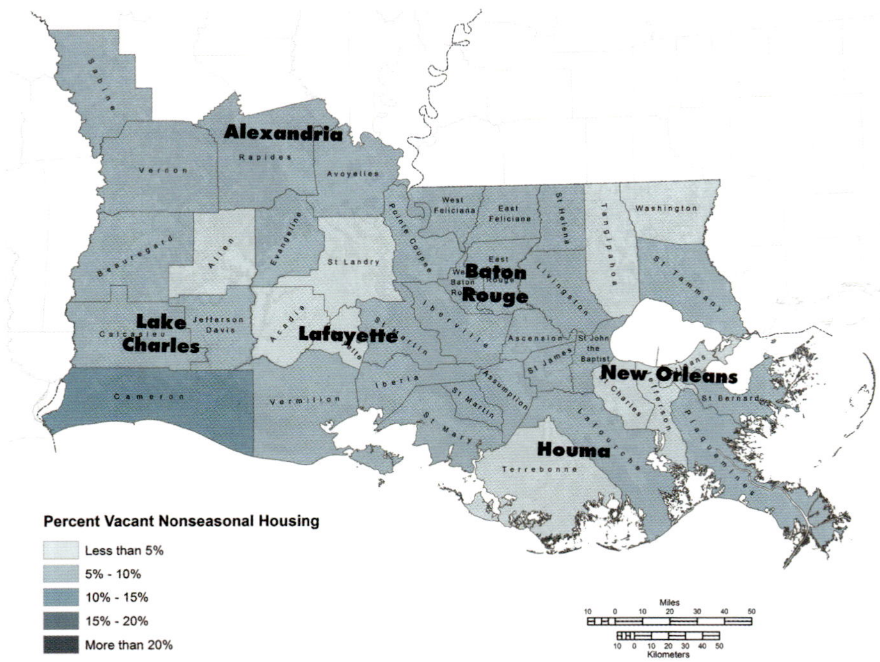

1950

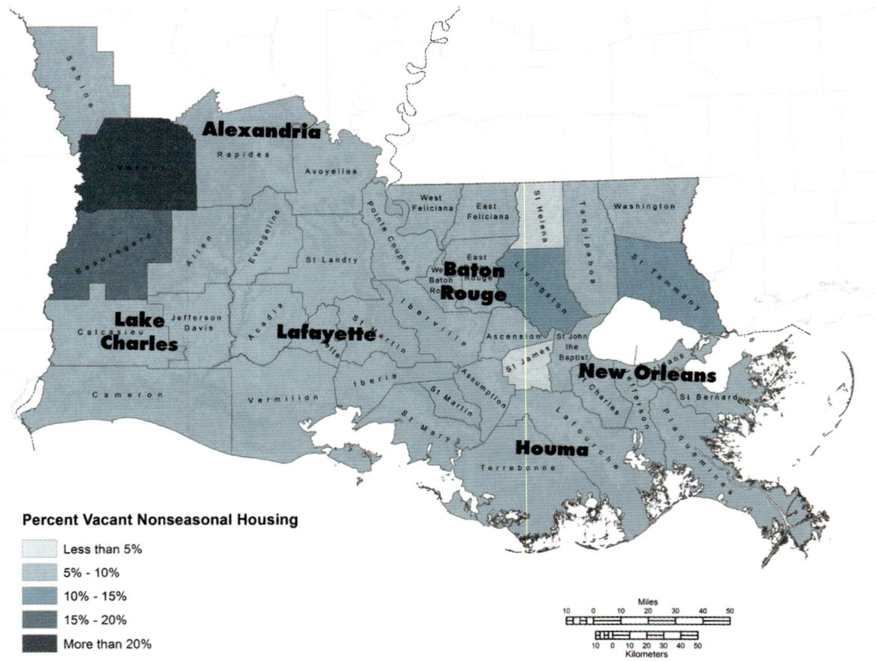

1960

One of the most common indicators of both community resilience and recovery after storm events and other shocks is the amount of vacant housing in a community or region. Vacant housing may indicate economic instability or population departure from an area.

Vacant housing may also indicate that there is a mismatch between anticipated and actual population levels (fig. 66). In the case of Fort Polk in 1960, a military base was not being utilized to its full extent. That situation was turned around within a decade, as the Vietnam War ramped up. In some cases, the impact of large hurricanes drives many residents out of a region, as we saw in 2010 following the aftermath of Hurricane Katrina. However, four decades earlier, in 1970, when the same area was dealing with the effects of Hurricanes Camille and Betsy, the region had extremely low home vacancy rates, suggesting a qualitative difference between community responses to these storm events.

Fig. 66. Vacant Housing in Coastal Shoreline and Coastal Watershed Parishes

Data Source. U.S. Census Bureau

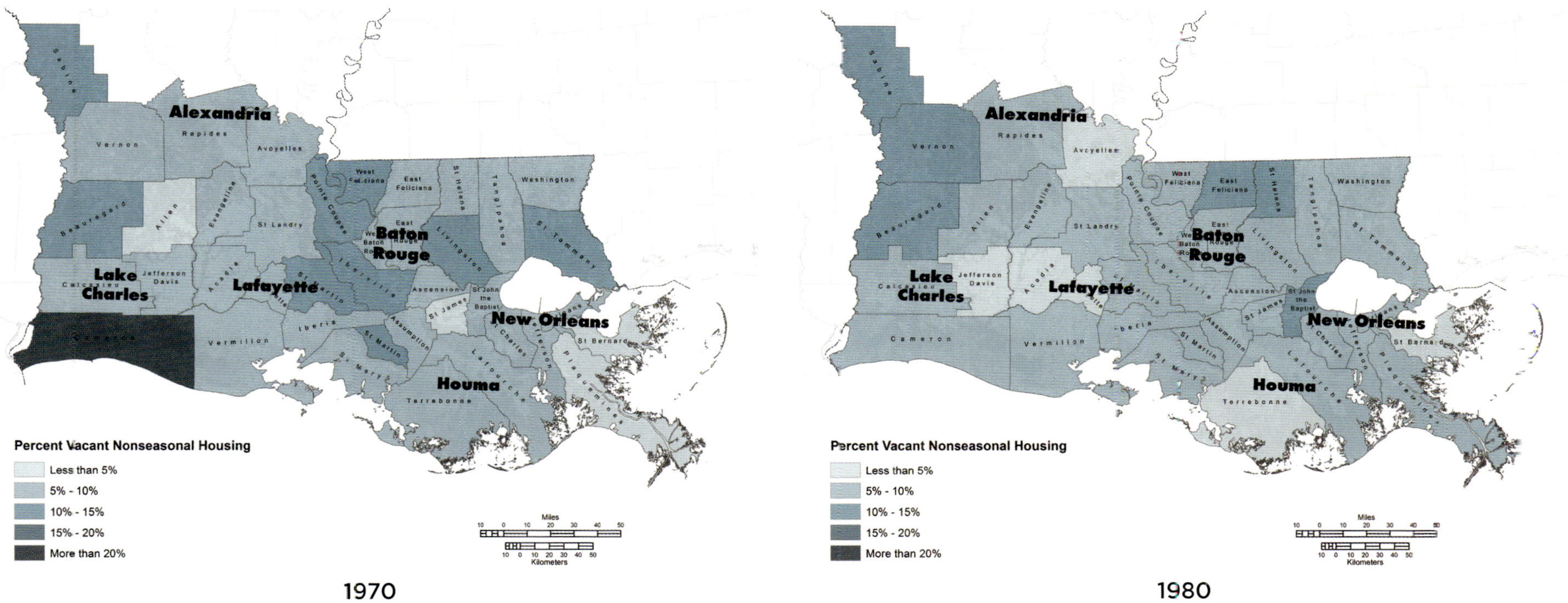

Fig. 66. Vacant Housing in Coastal Shoreline and Coastal Watershed Parishes

Data Source: U.S. Census Bureau

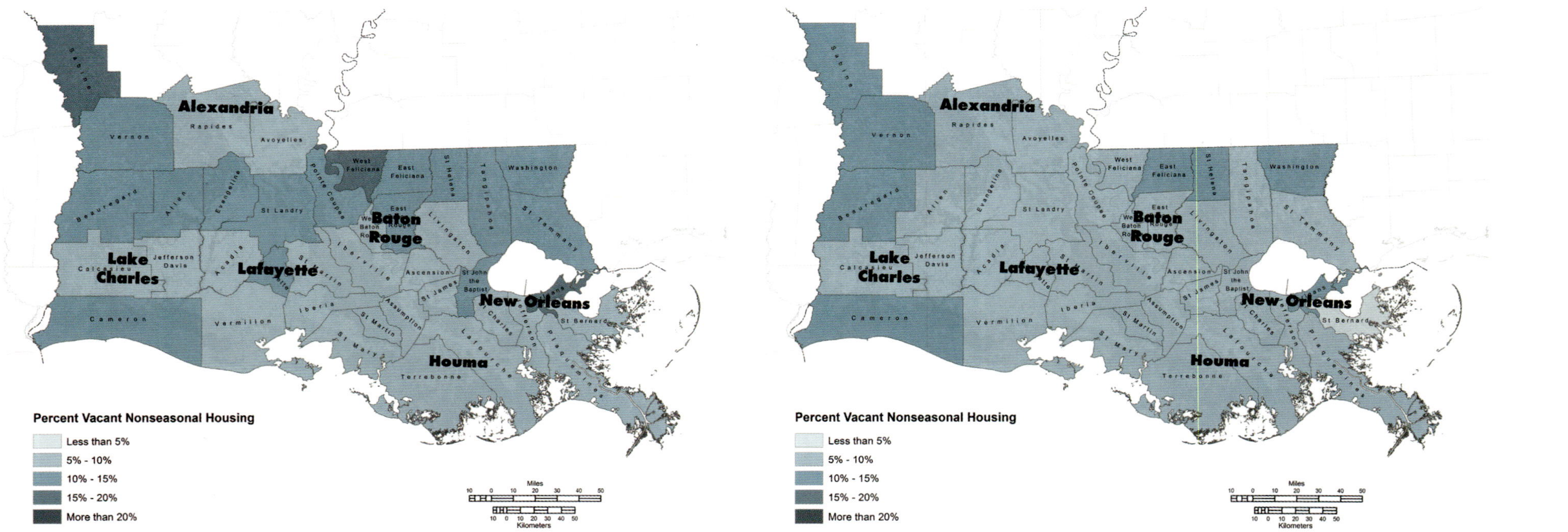

1990

2000

Fig. 66. Vacant Housing in Coastal Shoreline and Coastal Watershed Parishes

Data Source: U.S. Census Bureau

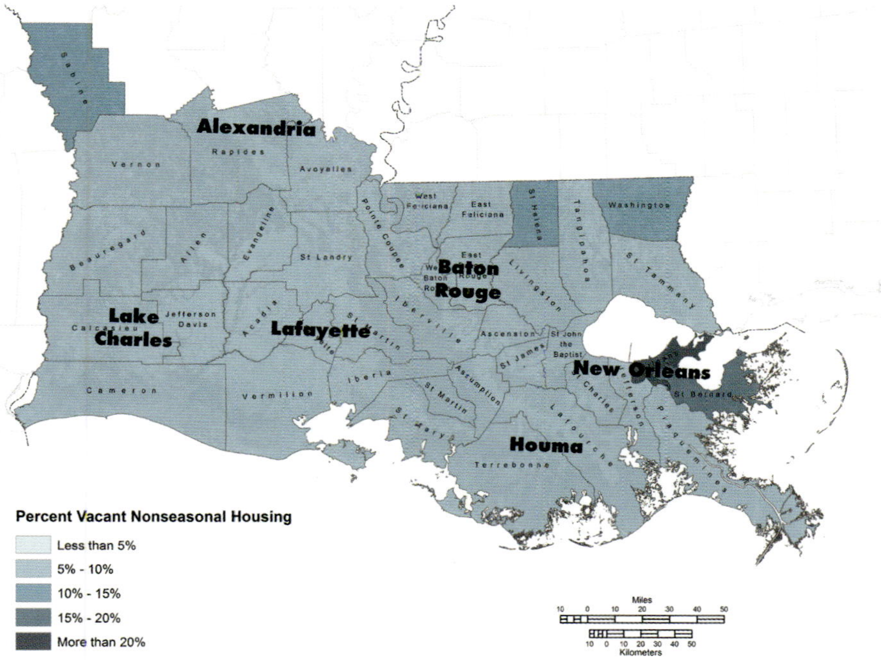

2010

Fig. 67. Regional Unemployment Clusters in Louisiana (1950–2010)

7 maps

Data Source: U.S. Census Bureau

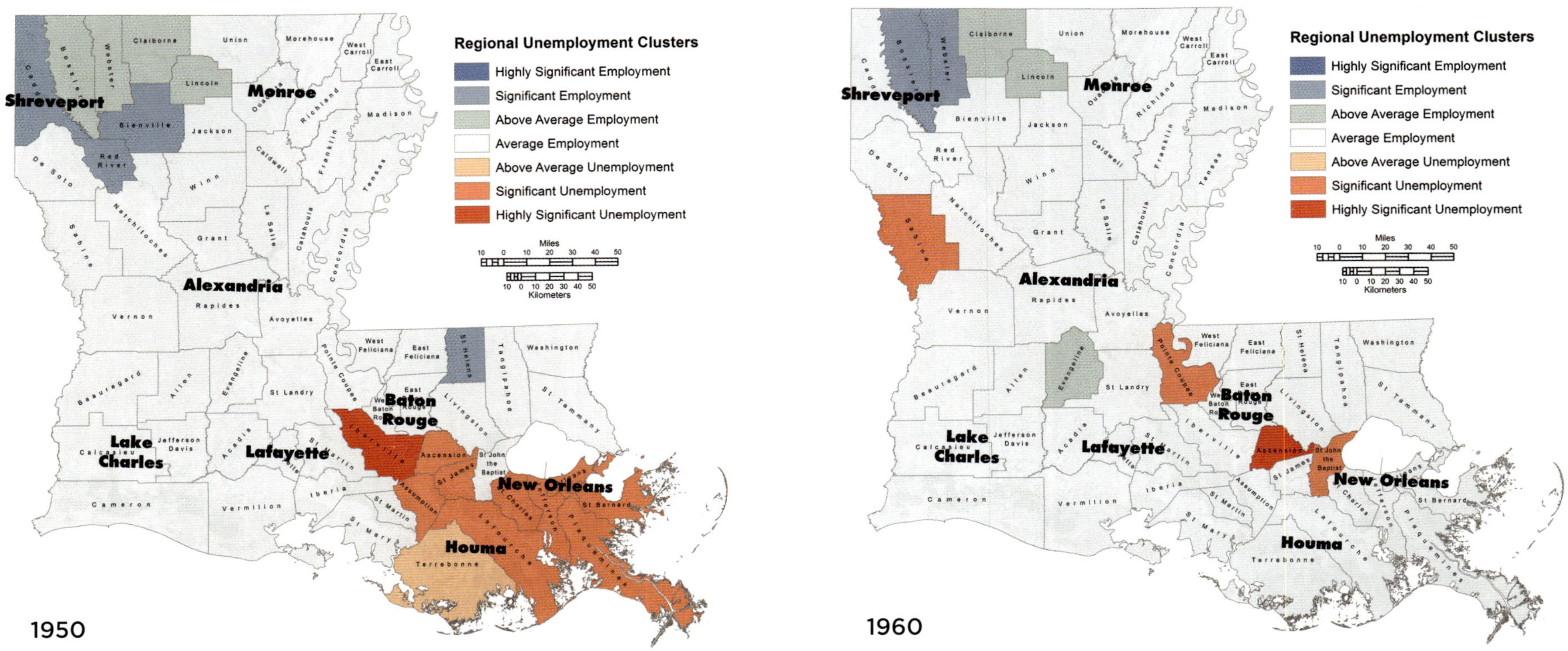

One final factor in population migration is employment (fig. 67). Access to employment opportunities directly influences regional migration patterns. Historically, the construction of the interstate highway system, as well as other state highway networks, has afforded residents the opportunity to live away from their employment sites, creating a regionalization of employment and unemployment. Analysis of regional unemployment clusters by decade reveals that, while there are certain regions where persistent unemployment is an issue, regional expansion and contraction of employment levels is a continuing and cyclic process. In coastal Louisiana, high levels of unemployment relative to the rest of the state were the pattern in the southeast area in 1950.

Fig. 67. Regional Unemployment Clusters in Louisiana (1950–2010)
Data Source: U.S. Census Bureau

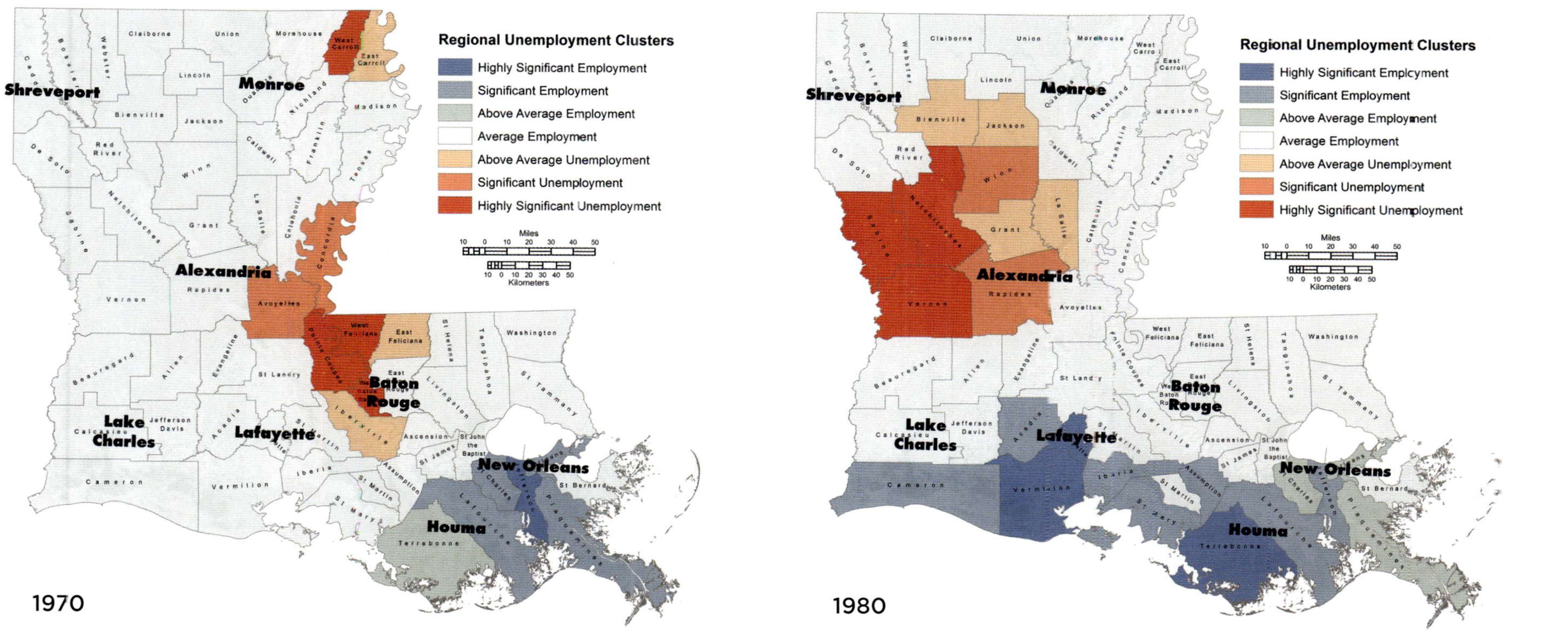

Subsequent decades saw significant growth in employment levels across the coast, particularly in 1970 and 1980 as the oil industry boomed in the region. Some of this growth was also likely the result of construction work required to assist the region in recovering from the large hurricane events of the 1960s. Following the oil bust of the 1980s and the subsequent shift to deepwater oil production, Lafourche Parish became a primary employment center. The average to low levels of unemployment through much of the coastal zone until the end of the twentieth century created a relatively stable economic environment, one of the key drivers of community resilience.

Fig. 67. Regional Unemployment Clusters in Louisiana (1950–2010)

Data Source: U.S. Census Bureau

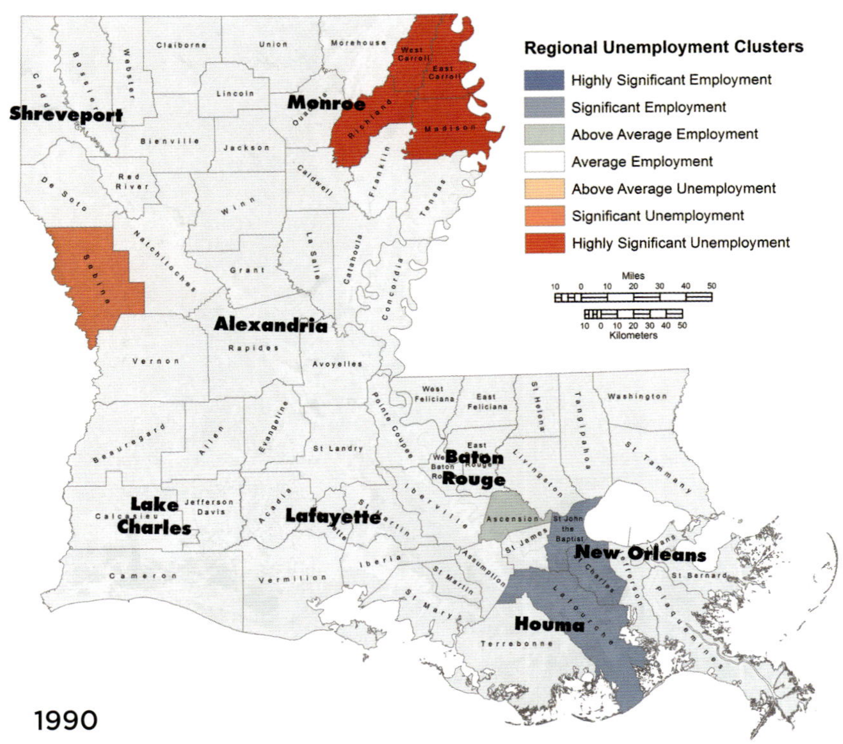

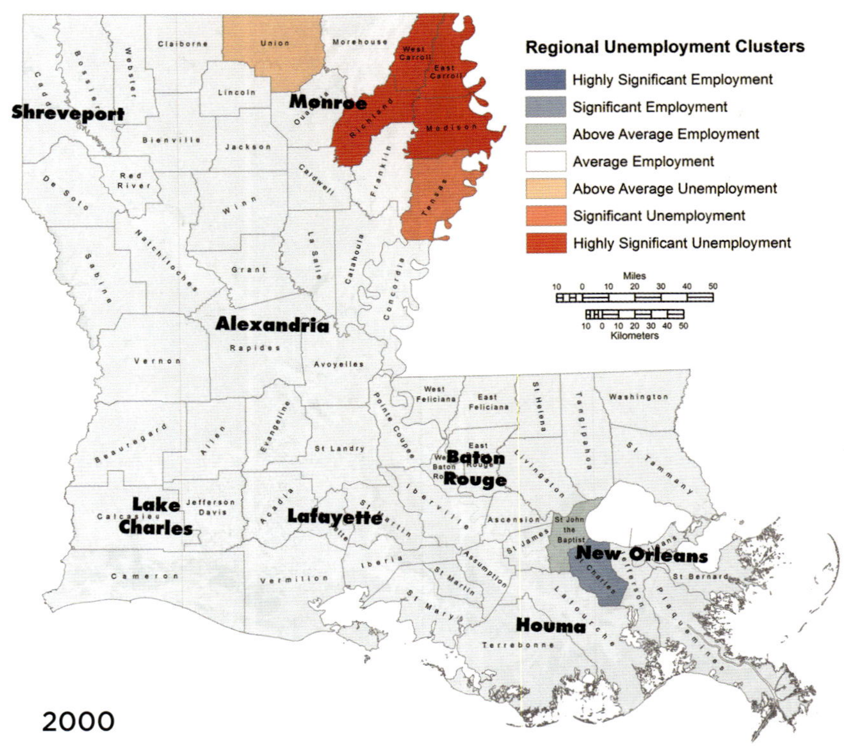

Fig. 67. Regional Unemployment Clusters in Louisiana (1950–2010)
Data Source: U.S. Census Bureau

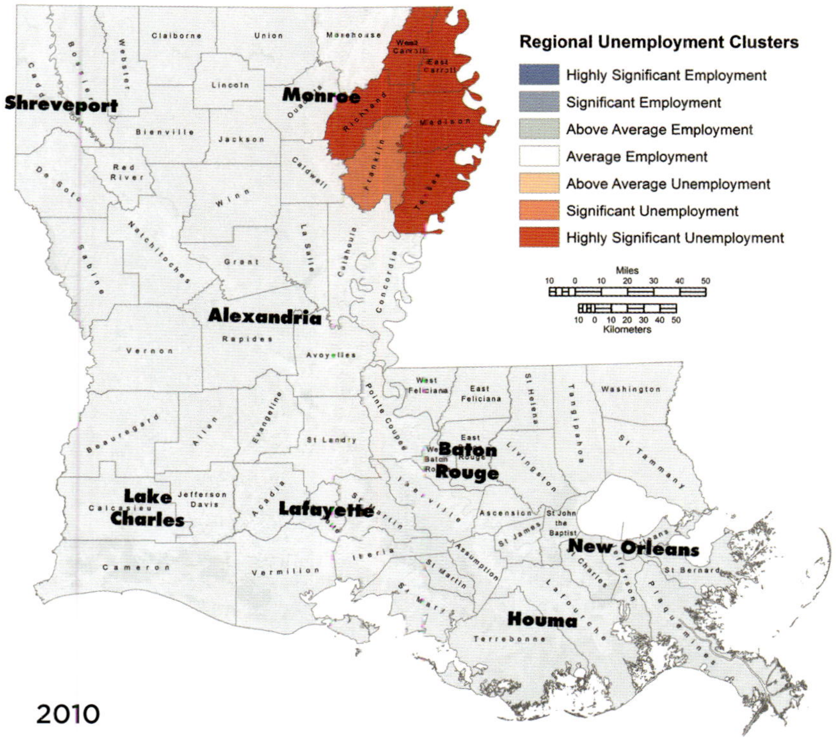

2010

MODIFYING RISK

While residents must constantly adapt to changing circumstances, these decisions are ultimately personal. When enough individuals make the same decision, communities change. Other types of adaptations are planned by larger governmental entities and designed to reduce risk for residents. The most common form of direct government intervention to try and reduce community risks and hazard exposure is the development of extensive coastal protection and restoration infrastructure.

Fig. 68. Structural Flood Protection

5 maps

Data Source: U.S. Army Corps of Engineers

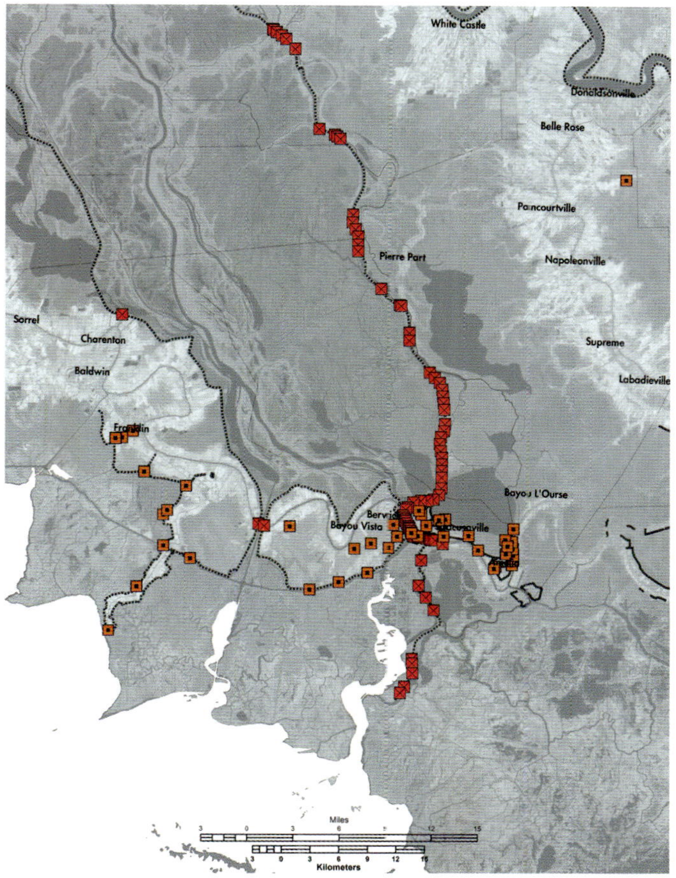

Atchafalaya (1)

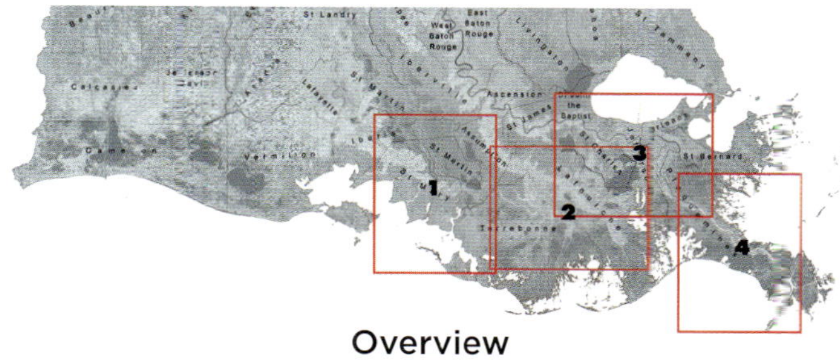

Overview

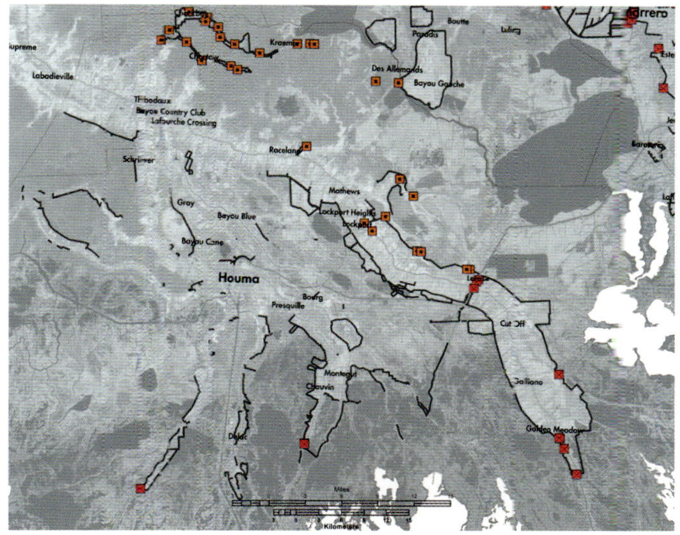

Houma (2)

A combination of population growth and limited high ground atop the state's natural levees has prompted the movement of populations into areas of low elevation where they are more vulnerable to flood events. In response, the federal government constructed an extensive network of river control and hurricane protection levees (fig. 68). While these levees have been shown to provide protection against most storm surges and river floods, they have had unintended consequences. Chief among these is enhanced subsidence in the leveed areas. This lowered elevation actually increases the risk of flooding in these protected areas, whether from large-scale events such as storm surge or from smaller-scale events such as extreme rainfall. For this reason, many urban areas within levee polders are still considered to be flood hazard areas.

Fig. 68. Structural Flood Protection

Data Source: U.S. Army Corps of Engineers

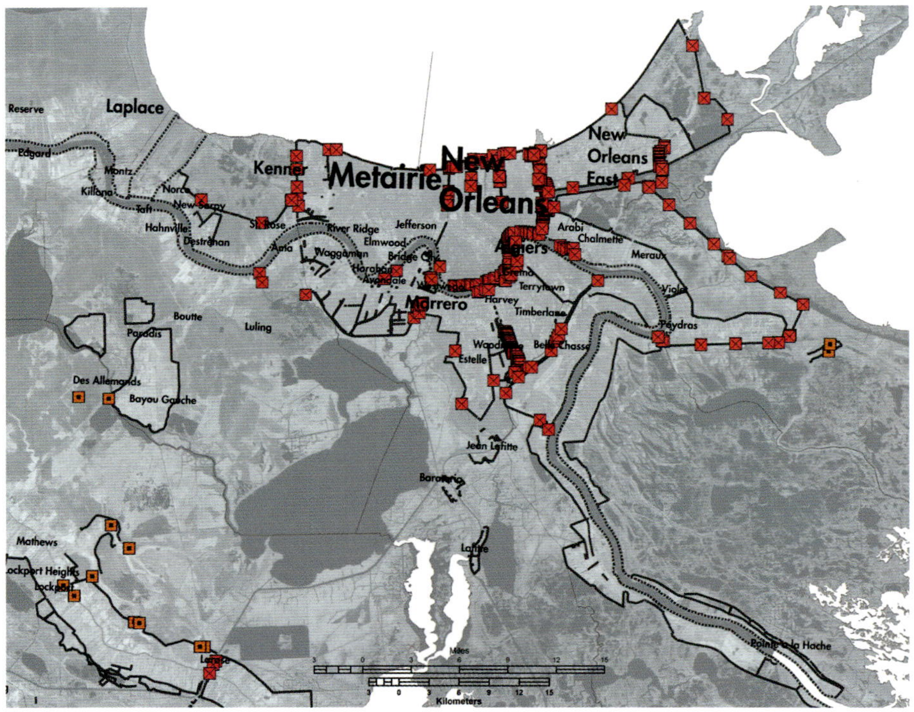

New Orleans (3)

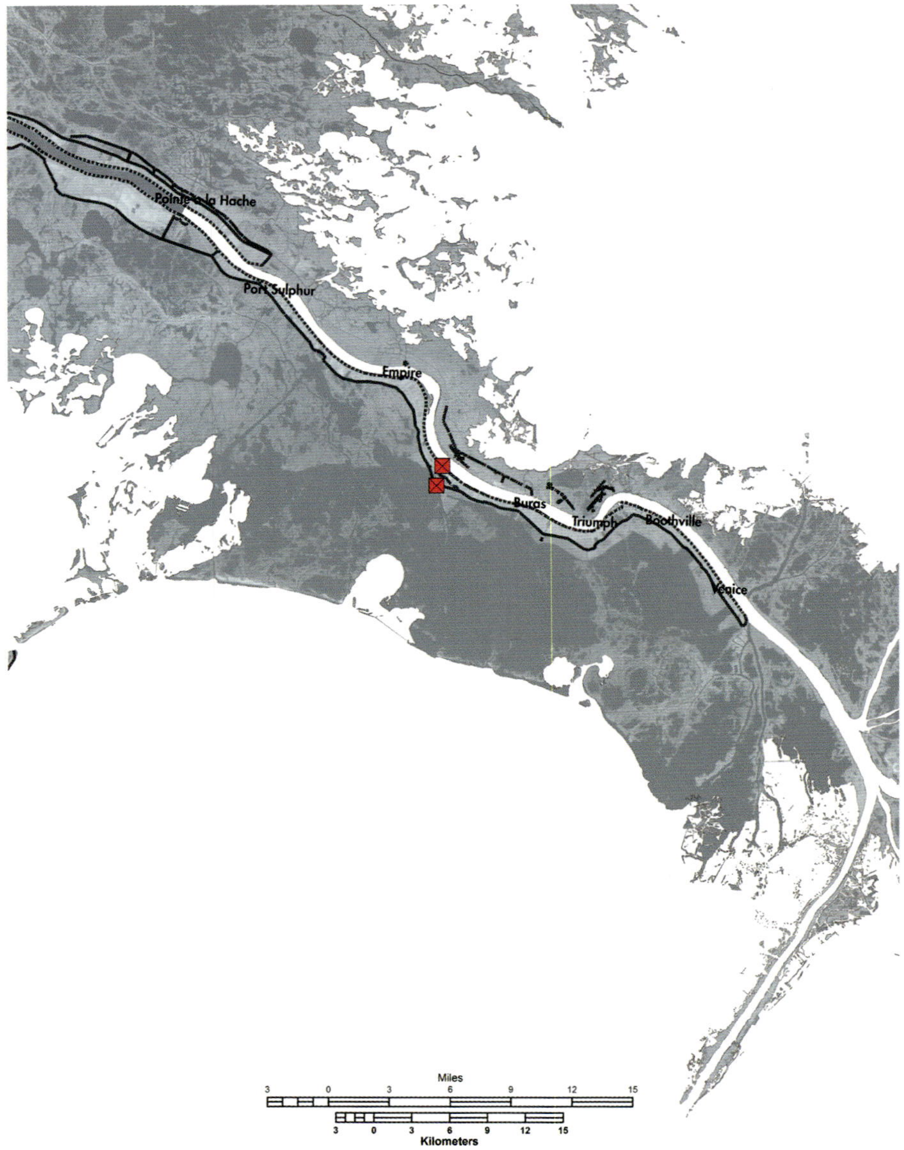

Plaquemines Parish (4)

Fig. 69. Coastal Protection Programs

7 maps

Data Source: Coastal Protection and Restoration Authority of Louisiana

Total Constructed Coastal Restoration Features

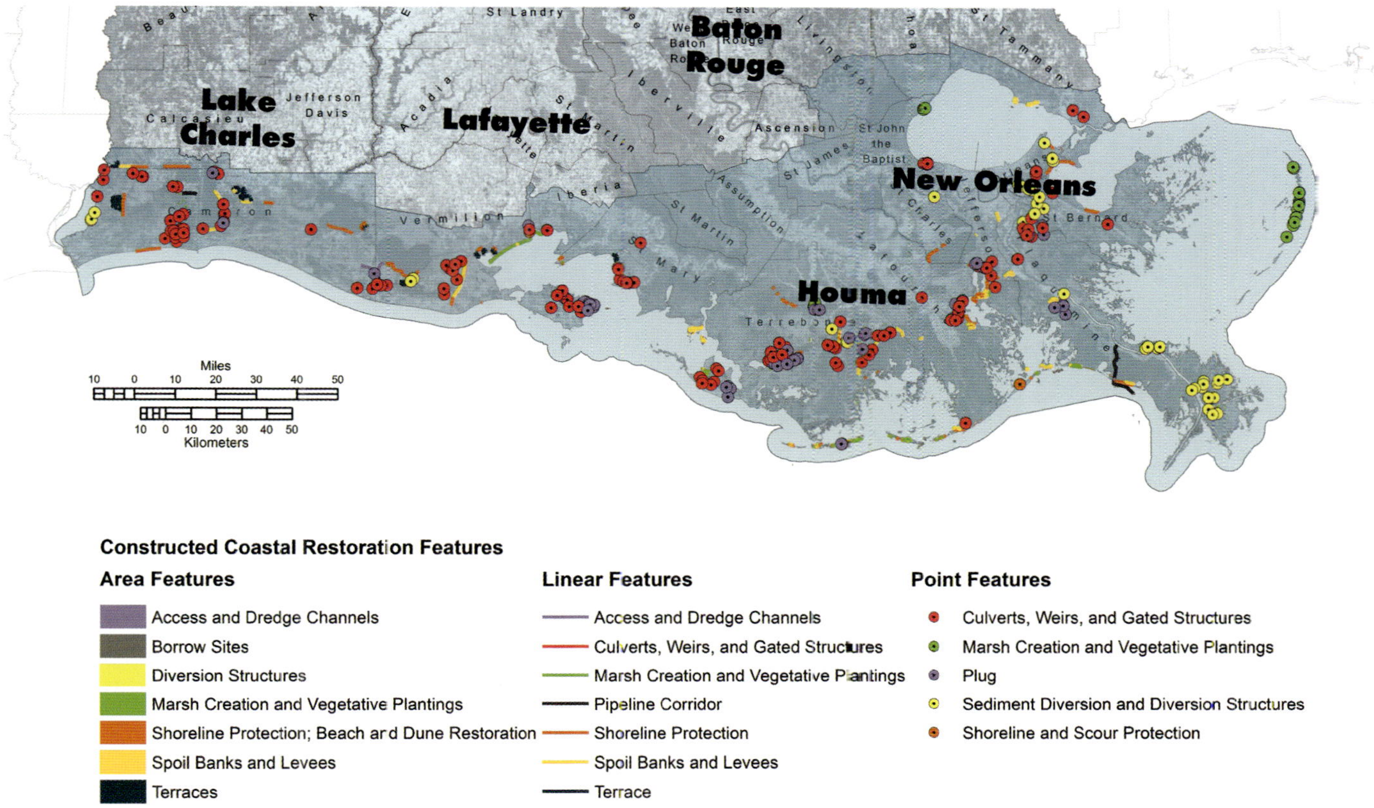

The modern levee construction that has largely reduced levels of river flooding has had other unintended consequences, including the loss of river sediment that has historically nourished the coastal wetlands. Through time, the loss of this sediment has caused wetlands to deteriorate and, in some cases, vanish entirely, converted to open water. The end of the twentieth century and the start of the twenty-first have seen the growth of a coastal restoration program in Louisiana with a goal of trying to halt the deterioration of the state's coastline (fig. 69). Numerous state and federal government programs have been developed to try to rebuild lost coastal land and the ecology of the region. In 1990, Congress passed the Coastal Wetlands Plan-

Fig. 69. Coastal Protection Programs

Data Source: Coastal Protection and Restoration Authority of Louisiana

Community Development Block Grant Program

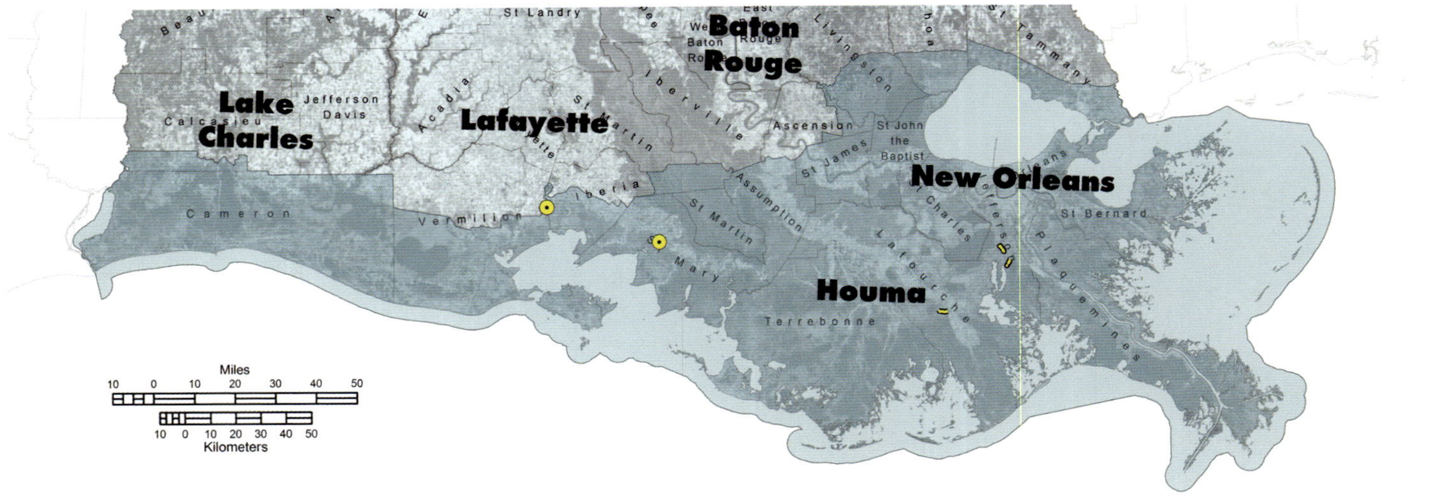

Coastal Impact Assistance Program

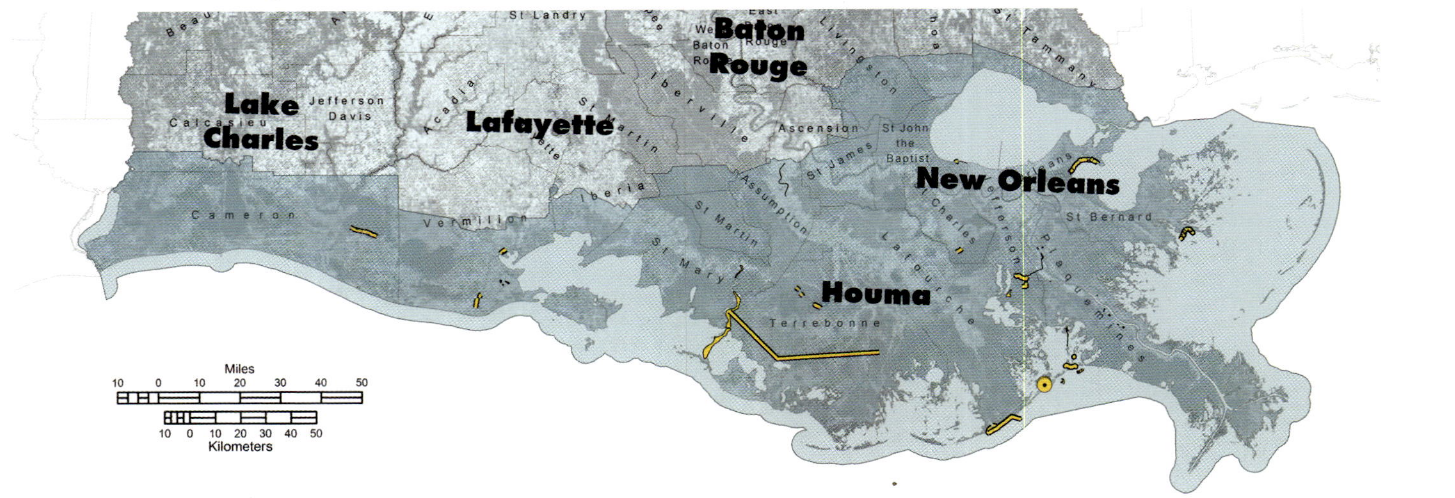

ning, Protection and Restoration Act (CWPPRA), the first federally mandated restoration effort aimed at Louisiana's coast. CWPPRA has brought an influx of federal funds into Louisiana earmarked specifically for coastal restoration projects. Other programs such as the Coastal Impact Assistance Program (CIAP) and the Water Resources Development Act (WRDA) followed, as did a number of projects funded solely by the State of Louisiana. CWPPRA and each of the subsequent programs have tended to focus on very specific geographical areas. Nevertheless, this influx of funding for these types of projects represents a recognition of the coastal environment as both a valuable natural resource and a valuable source of protection for coastal communities.

Fig. 69. Coastal Protection Programs

Data Source: Coastal Protection and Restoration Authority of Louisiana

Coastal Wetlands Planning, Protection and Restoration Act

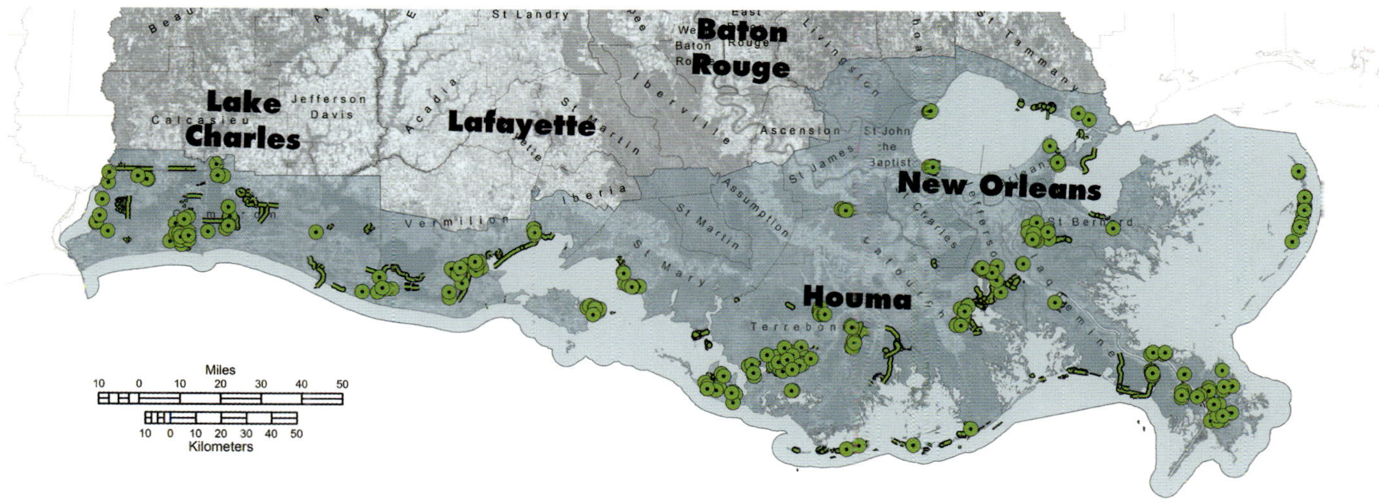

State Projects

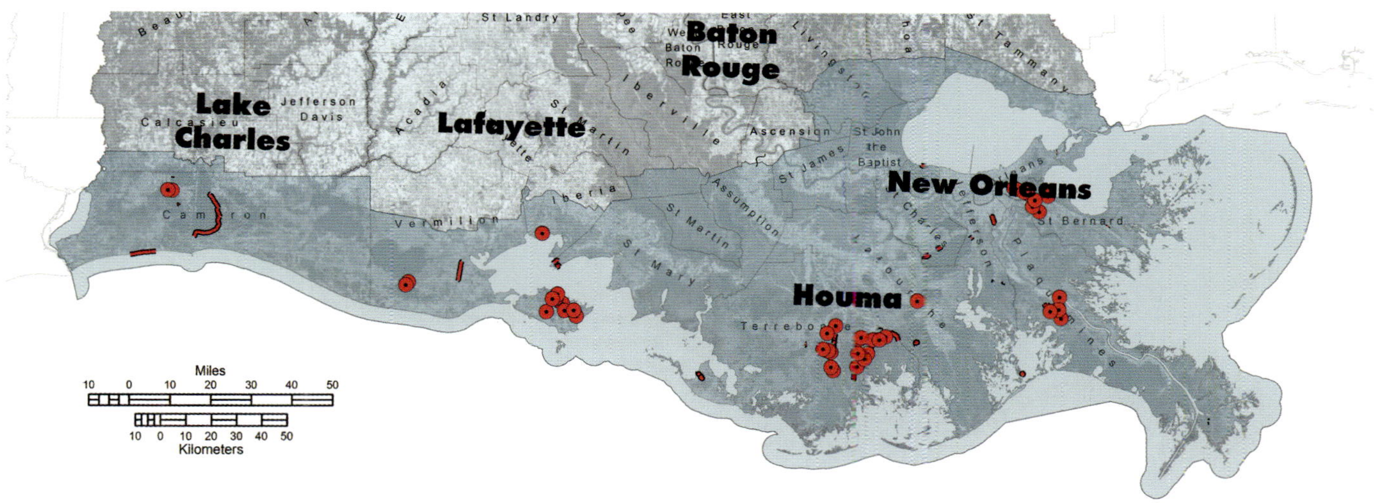

Constructed Coastal Restoration Features

Area Features

- Access and Dredge Channels
- Borrow Sites
- Diversion Structures
- Marsh Creation and Vegetative Plantings
- Shoreline Protection; Beach and Dune Restoration
- Spoil Banks and Levees
- Terraces

Linear Features

- Access and Dredge Channels
- Culverts, Weirs, and Gated Structures
- Marsh Creation and Vegetative Plantings
- Pipeline Corridor
- Shoreline Protection
- Spoil Banks and Levees
- Terrace

Point Features

- Culverts, Weirs, and Gated Structures
- Marsh Creation and Vegetative Plantings
- Plug
- Sediment Diversion and Diversion Structures
- Shoreline and Scour Protection

Fig. 69. Coastal Protection Programs

Data Source: Coastal Protection and Restoration Authority of Louisiana

Hurricane and Storm Damage Risk Reduction System

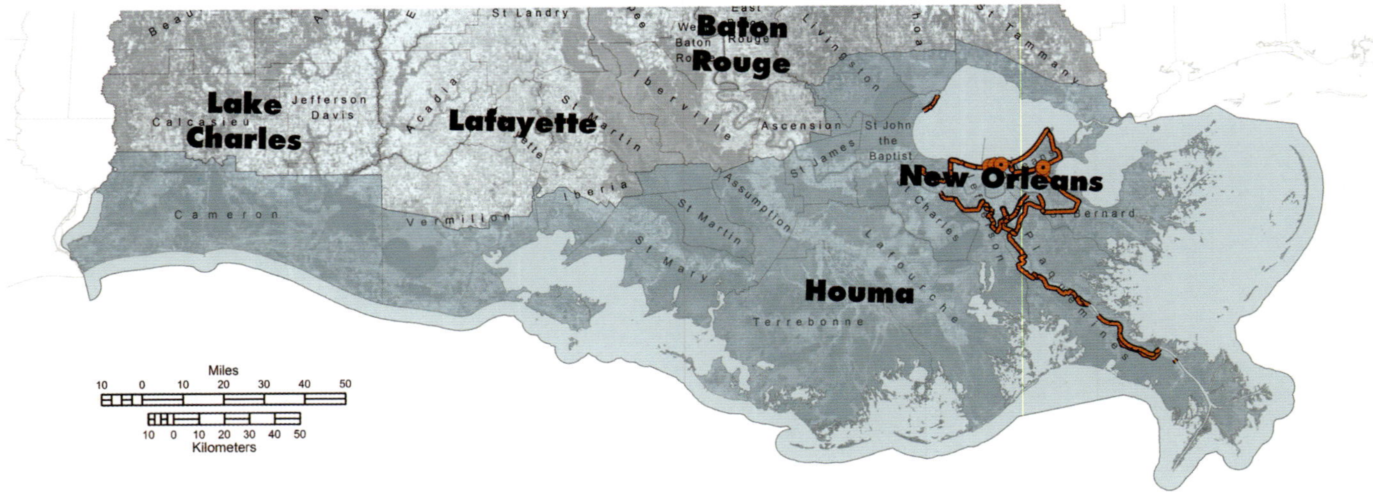

Water Resources Development Act

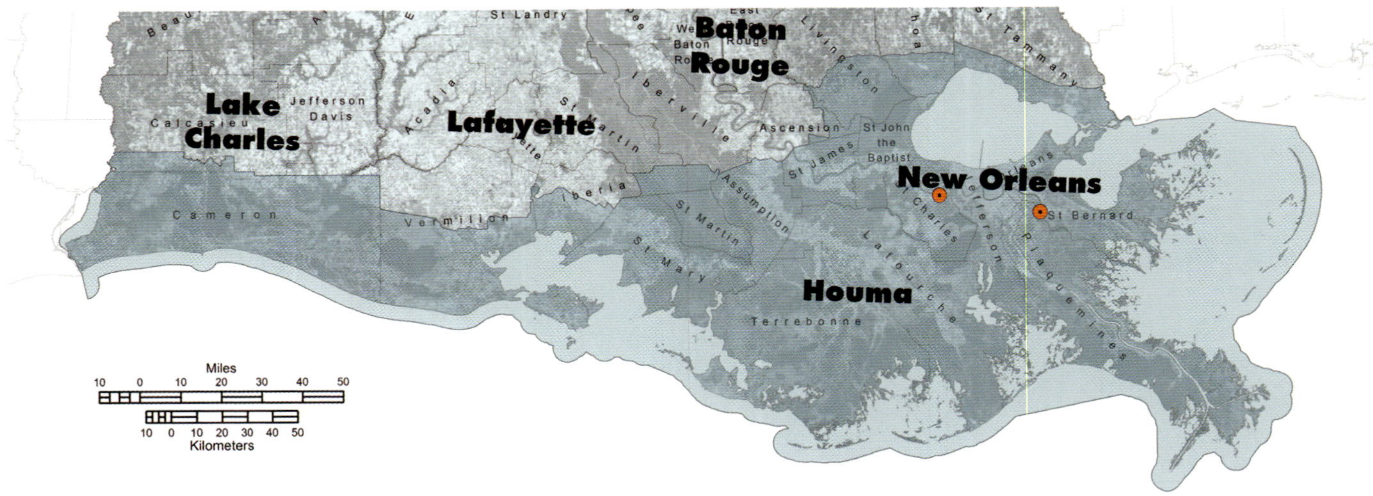

Constructed Coastal Restoration Features

Area Features

- Access and Dredge Channels
- Borrow Sites
- Diversion Structures
- Marsh Creation and Vegetative Plantings
- Shoreline Protection; Beach and Dune Restoration
- Spoil Banks and Levees
- Terraces

Linear Features

- Access and Dredge Channels
- Culverts, Weirs, and Gated Structures
- Marsh Creation and Vegetative Plantings
- Pipeline Corridor
- Shoreline Protection
- Spoil Banks and Levees
- Terrace

Point Features

- Culverts, Weirs, and Gated Structures
- Marsh Creation and Vegetative Plantings
- Plug
- Sediment Diversion and Diversion Structures
- Shoreline and Scour Protection

CHAPTER 4

TRANSFORMATION OF THE STATUS QUO

Adapting to the Changing Coastal Landscape

CHANGE IN THE LANDSCAPE of coastal Louisiana is inevitable. History has shown that this region is, indeed, a perilous environment in which coastal residents are at constant environmental and economic risk. And while society has persisted, adapted, and, in some cases, even thrived in this dynamic environment, the rate of environmental change is expected to increase in the coming years, placing many more coastal residents in harm's way. The looming specter of climate change will challenge the ingenuity of Louisianans, both residents and their elected officials, in new and powerful ways.

Coastal land loss, despite the best efforts of residents, scientist, and planners, is expected to continue. Although deterioration of the coastline can be mitigated to some extent by human intervention, the continued loss of the wetlands in the coming years will potentially result in adverse impacts to landscape structure, hurricane and storm protection, and valuable ecosystem services that coastal communities rely on for their well-being (Couvillion et al. 2013). If no additional actions are taken to slow the rate of loss, the impacts will be devastating. The removal of the wetlands, an invaluable buffer against hurricanes and storm surge, places coastal communities ever closer to open water and the risks inherent to such proximity.

Rising sea levels will only compound this issue. As sea levels rise, coastal shorelines retreat and low-lying coastal areas are inundated more frequently by the waters of the Gulf of Mexico. The greatest potential impacts of sea-level rise in coastal Louisiana occur in locations where physical vulnerability coincides with concentrations of economically valuable infrastructure. Communities located along the Barataria and Terrebonne basins, for example, are extremely vulnerable to future hazard events due to the combination of high local rates of relative sea-level rise, significant shoreline erosion and wave energy, and a concentration of economically valuable infrastructure (Thatcher et al. 2013). Adding to the vulnerability of these communities is the high density of population development atop the narrow threads of high ground and natural levees in the region.

Past efforts to provide protection, in the form of large-scale levee projects, have proven successful when examined historically. The risk of river flooding—a constant threat in coastal Louisiana during both the storm season and the northern spring thaws, when melting snows in the northern and central United States send large volumes of water down the large continental river systems—has been significantly reduced. The levee systems have provided a buffer against tropical storm surges time and again. Over time, the system of structural protection within coastal Louisiana has, on balance, protected a great majority of residents from physical and economic harm.

And yet the system is not infallible. Levees have at times proven inadequate, both in terms of design and construction. Additionally, the presence of levees often encourages people to move into and develop areas that are inherently more hazard-prone. As population densities are maximized atop the natural levees, populations migrate into lower-lying areas off of these levees where the flood risks are more significant. This problem is compounded as this development often accelerates the rates of subsidence in the leveed areas and beneath the levees themselves. As a result, more and more populations are at risk of flooding in seemingly well-protected areas.

For coastal Louisiana to remain vibrant and for the working coast to continue to produce the resources that local communities, the state, and the nation rely on, residents, scientists, and planners must continue to adapt to these changing conditions. Adaptation is an ongoing process building off of previously successful adaptations and discarding ill-conceived, maladaptive practices.

To effectively prepare for and adapt to unrelenting environmental changes, it is necessary to take an integrated approach to coastal protection and restoration. Decades of earlier planning efforts have focused on either coastal protection or coastal restoration or were focused on geographically limited locations. Ultimately, these piecemeal efforts have had limited success and have not resulted in a net gain of wetlands (Peyronnin et al. 2013). New adaptive measures are beginning to acknowledge the systematic complexity of interrelated issues in Louisiana's coastal zone and to develop more coordinated, integrative frameworks. This multiple-lines-of-defense strategy incorporates a broad suite of structural, nonstructural, and coastal restoration features, including the Gulf of Mexico shelf, the barrier islands, the sounds, marsh land bridges, natural ridges, manmade ridges, floodgates, flood levees, pump stations, home and building elevations, and evacuation routes (Lopez 2009). This integrative strategy should also include broader policy-related adaptations, including targeted land development, coordinated transportation investments, stronger building codes, and a strategic approach to land regulation (CPEX 2012). The incorporation of resilience-based planning strategies will reduce the types of maladaptive development that have, historically, placed populations in harm's way.

Ultimately, the adaptive capacity of coastal communities is highly dependent on the ability of residents to live with the environment. The environment has long been a source of sustenance for coastal Louisiana, and it has also protected residents from the deleterious effects of coastal hazards and external environmental shocks for as long as humans have settled the region. While large-scale structural protection measures have largely supplanted the role of ecosystem-based protection measures in the minds of coastal residents and planners alike, an ecology-based approach remains a viable means to help protect residents as well as the protection infrastructure that they depend on. Any future adaptive strategy must therefore provide alternative ecosystem-based, nonstructural means to enhance protection in leveed areas and to provide protection to those outside them.

Soft engineering of Louisiana's coastline using a systemic, ecology-based approach is a vital tool to support coastal communities and their continuing relationship with the ecosystem. Building mangrove forests, restoring broad swaths of coastal marshes, protecting coastal forests, and constructing oyster reefs in offshore areas are all valuable alternative means of coastal protection. These types of bioshields can supplement or replace levees and storm surge barriers by physically slowing the progress of water from surges (Osti et al. 2009).

While large-scale integration of coastal restoration, coastal protection, and land-use regulation is central to the success of any future adaptation strategy to build resilience in coastal communities, this does not guarantee that these communities will become more resilient. Resilience cannot be imposed on communities in a top-down manner. The final lines of defense for many coastal communities, and those that the concept of community resilience may ultimately rely on more than any other, are community- and site-specific (CPEX 2012). Scientists and planners may address many of the larger, more global issues facing communities, but it is the small-scale, local adaptations that provide residents with an enhanced sense of place that may serve in many ways to vitalize, or revitalize, communities.

FUTURE RISK

Despite the best efforts of residents, planners, and government officials, the communities of coastal Louisiana still face unprecedented risks in the twenty-first century. The impacts of climate change, a result of both natural climate cycles and human intervention, are anticipated to challenge coastal residents in numerous ways.

Fig. 70. Area at Risk from Future Storm Surges

Data Source: Coastal Protection and Restoration Authority of Louisiana

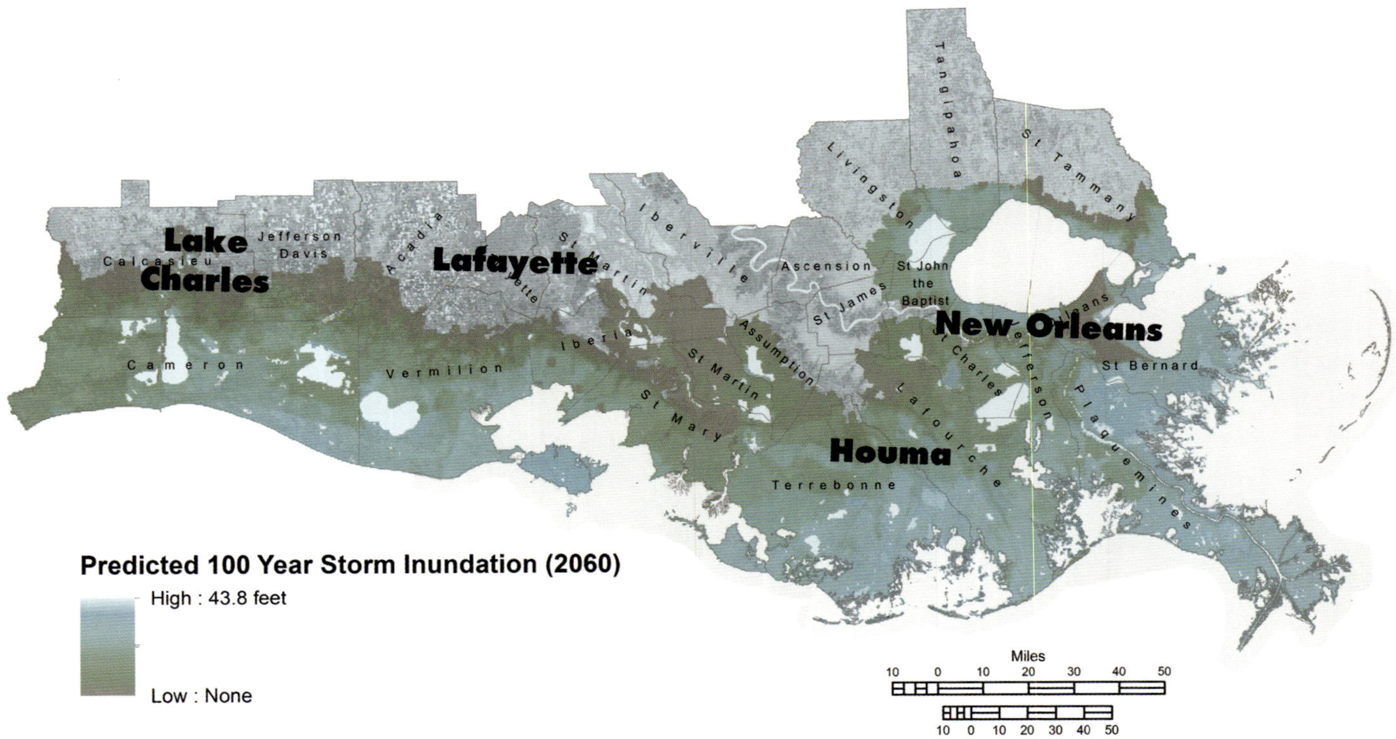

Relative sea-level rise, the combination of anticipated sea-level rise and the subsidence of coastal land, has the potential to affect coastal communities as never before (fig. 70). There is a great deal of uncertainty as to the degree of sea-level rise that the future will bring, but it is inevitable that coastal communities will face heightened levels of risk in the future.

Fig. 71. Area at Risk from Future Land Loss

Data Source: Coastal Protection and Restoration Authority of Louisiana

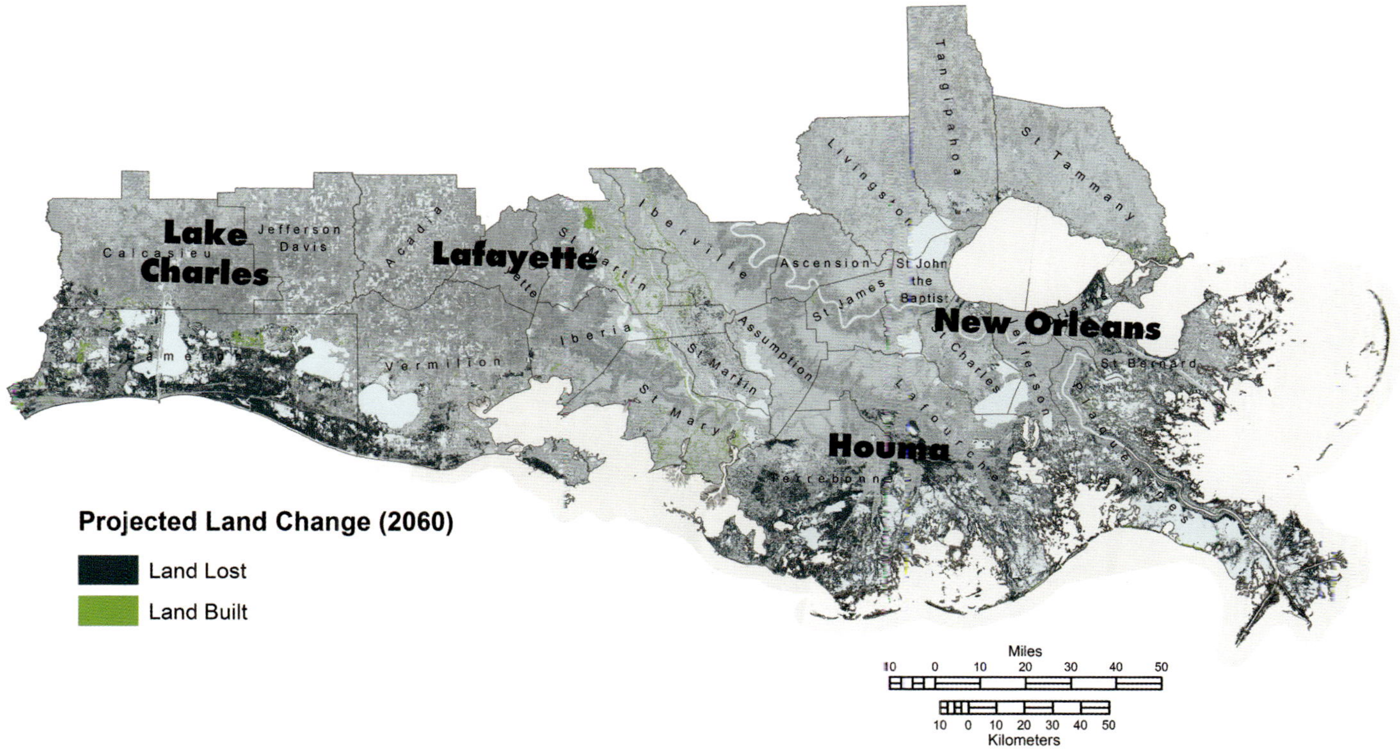

The same applies to continuing levels of coastal land loss. If no action is taken to protect and rebuild Louisiana's wetlands, coastal land loss will continue to be a serious issue for coastal residents. As this line of coastal defense continues to erode, the land-water line will move closer and closer to some of the state's largest and most populous communities, including Houma, Lake Charles, and New Orleans. If no additional actions are taken to halt or reduce the loss, the last vestiges of this valuable coastal buffer could wash away within the next half century (fig. 71).

FUTURE ADAPTATION AND INNOVATIVE PRACTICES

What is required if coastal Louisiana is to persist, even thrive, is a continued push to adapt. Innovative techniques must be applied at all scales, from the local to the global.

Fig. 72. Selected Community-Scale Adaptation Measures

6 maps

Data Sources: American Society of Landscape Architects; Center for Planning Excellence; Groundwork New Orleans; Louisiana Direct Seafood; U.S. Environmental Protection Agency

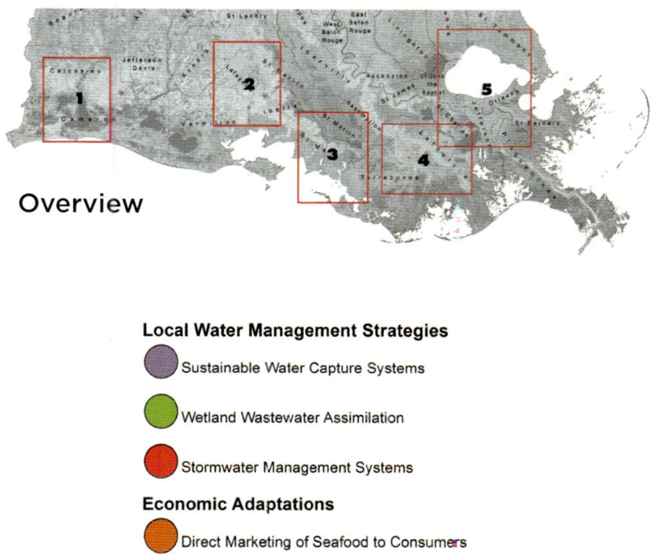

Overview

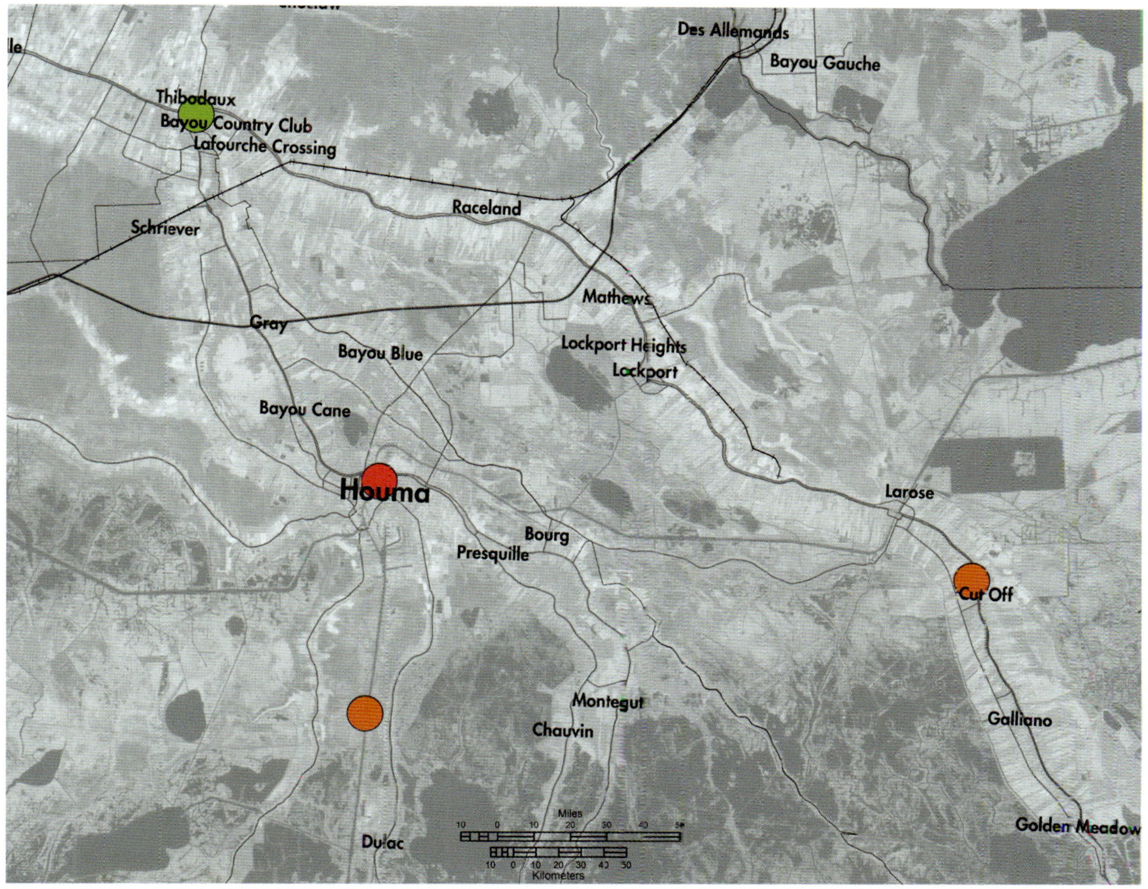

Houma (4)

Innovations are already occurring at the local level (fig. 72). Many of these are designed to help communities better manage storm water and municipal wastewater, by reducing levels of impervious land in urban centers through the development of greenways and water gardens. Other communities are using municipal effluent to nourish local wetlands. While these local innovations cannot eliminate the impacts of extreme events such as coastal storm surges, they can reduce the impacts of smaller-scale extreme weather and rainfall events. Other local adaptations are economic in nature. This includes a shift to the direct marketing of seafood to consumers, a practice that allows fishermen more flexibility and opportunities for economic growth.

Fig. 72. Selected Community-Scale Adaptation Measures

Data Sources: American Society of Landscape Architects; Center for Planning Excellence; Groundwork New Orleans; Louisiana Direct Seafood; U.S. Environmental Protection Agency

Local Water Management Strategies

Sustainable Water Capture Systems

Wetland Wastewater Assimilation

Stormwater Management Systems

Economic Adaptations

Direct Marketing of Seafood to Consumers

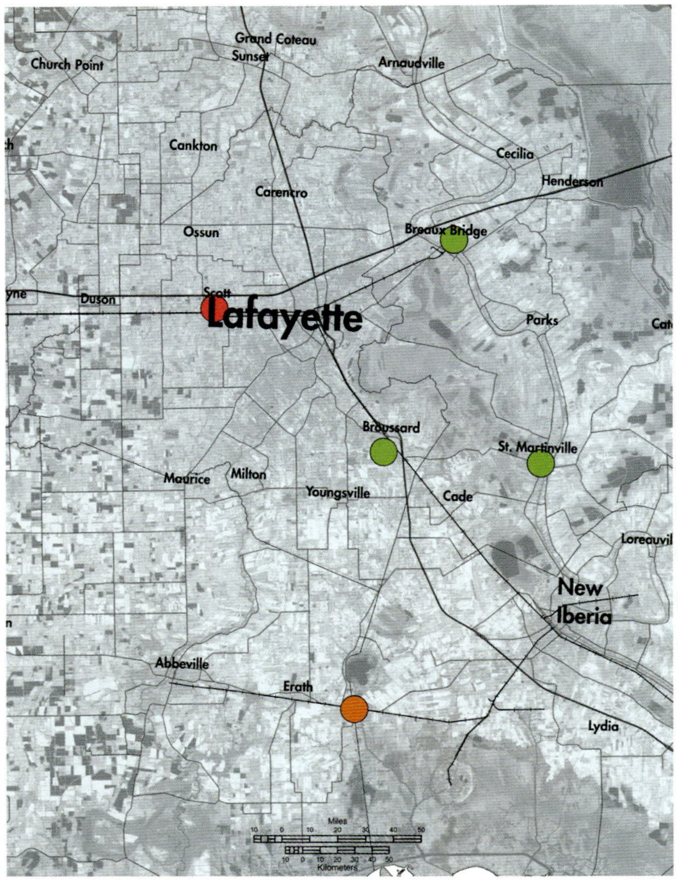

Lafayette (2)

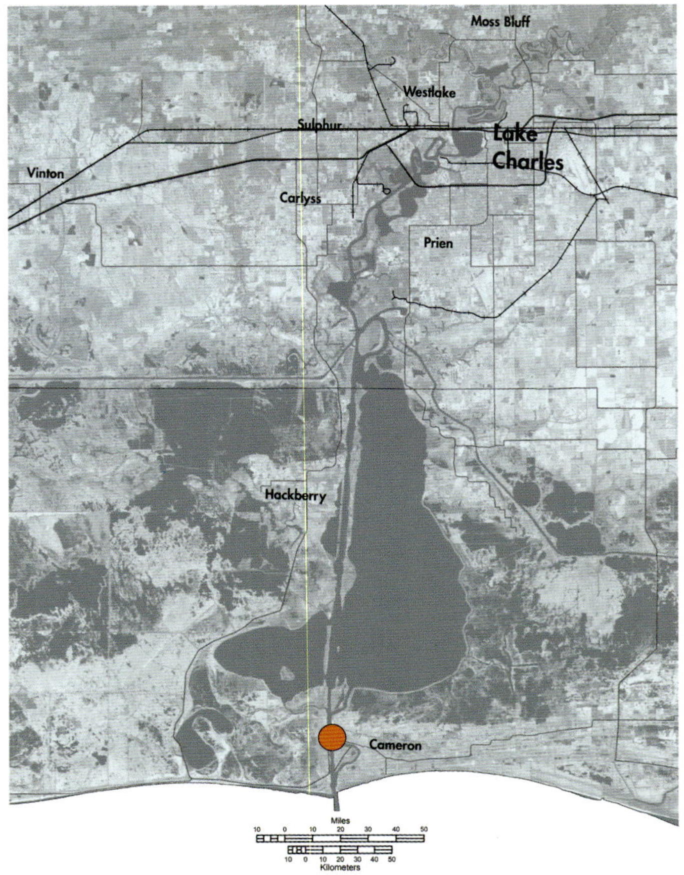

Lake Charles (1)

Fig. 72. Selected Community-Scale Adaptation Measures

Data Sources: American Society of Landscape Architects; Center for Planning Excellence; Groundwork New Orleans; Louisiana Direct Seafood; U.S. Environmental Protection Agency

Local Water Management Strategies

- Sustainable Water Capture Systems
- Wetland Wastewater Assimilation
- Stormwater Management Systems

Economic Adaptations

- Direct Marketing of Seafood to Consumers

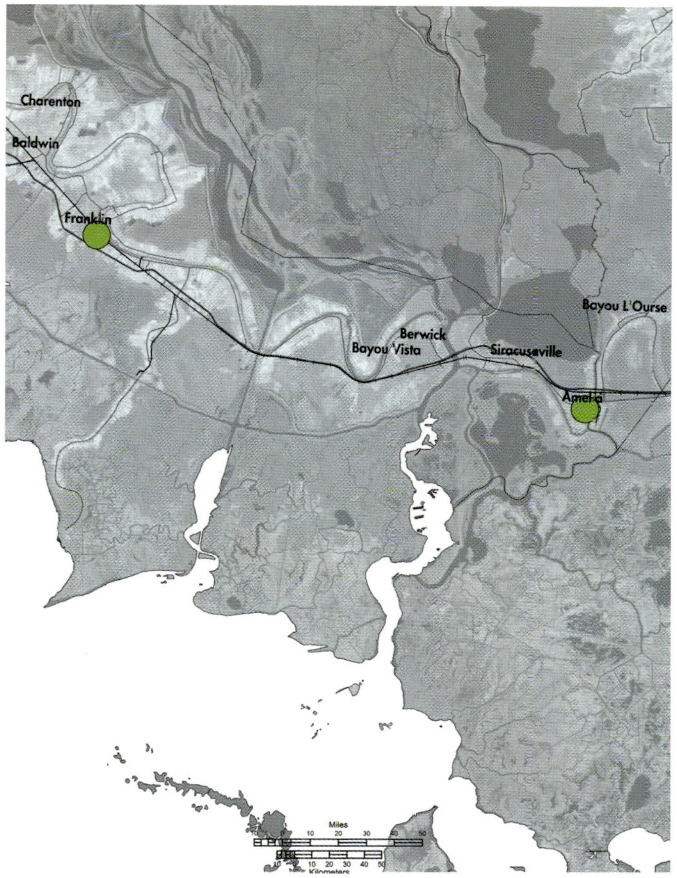

St. Mary Parish (3)

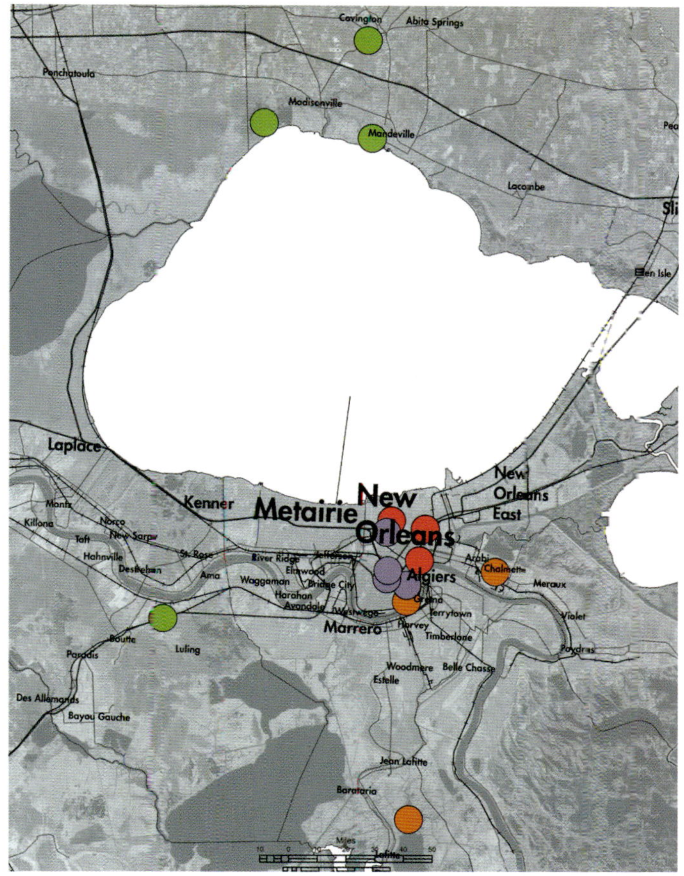

New Orleans (5)

Fig. 73. Number of Charter Boat Permits Issued

4 maps

Data Source: Louisiana Department of Wildlife and Fisheries

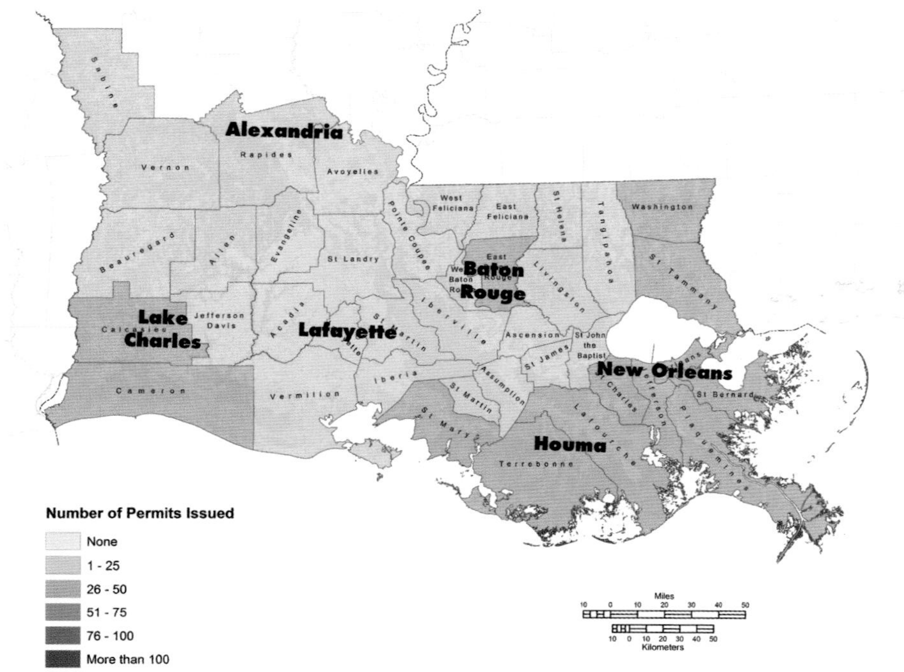

1995

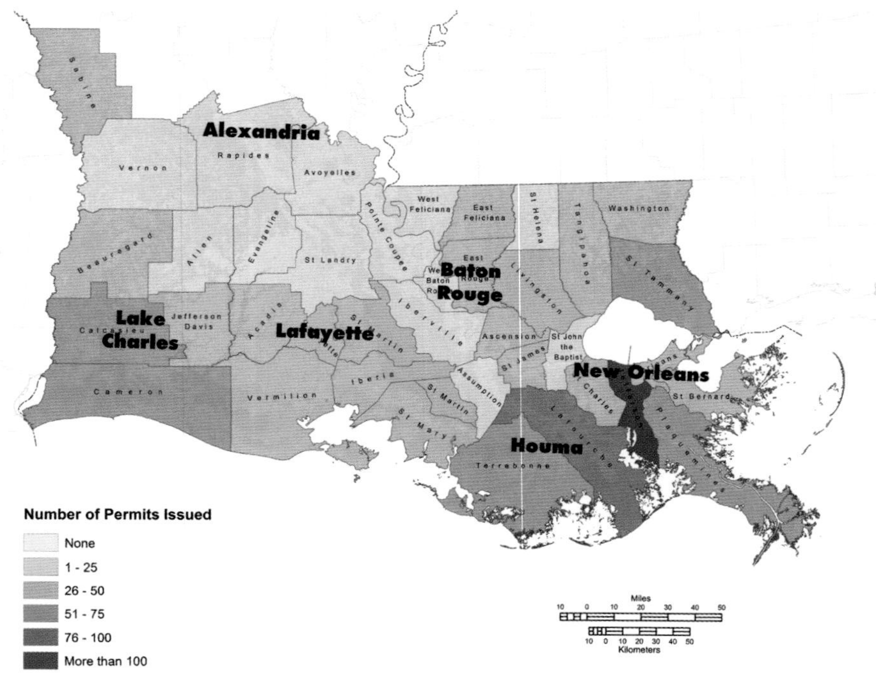

2000

Other adaptations are focused on the economic stability of residents who rely on the environment for employment and well-being. Some coastal residents have turned to ecotourism, organizing activities like swamp tours, bird-watching, pelagic bird tours, and deep-sea fishing excursions. The number of charter boat licenses has climbed steadily since 1995, suggesting that more and more coastal residents are engaging in the tourism industry, whether through ecotourism or fishing excursions, or a combination of both (fig. 73). The highest levels of growth in charter boat permits are apparent near New Orleans and the broader southeast Louisiana region.

Local adaptations like these enable communities to build on and develop their unique sense of place and make residents feel more invested in their communities. This sense of community is one of the intangible aspects that is key to the growth and development of inherent resilience.

Fig. 73. Number of Charter Boat Permits Issued

Data Source: Louisiana Department of Wildlife and Fisheries

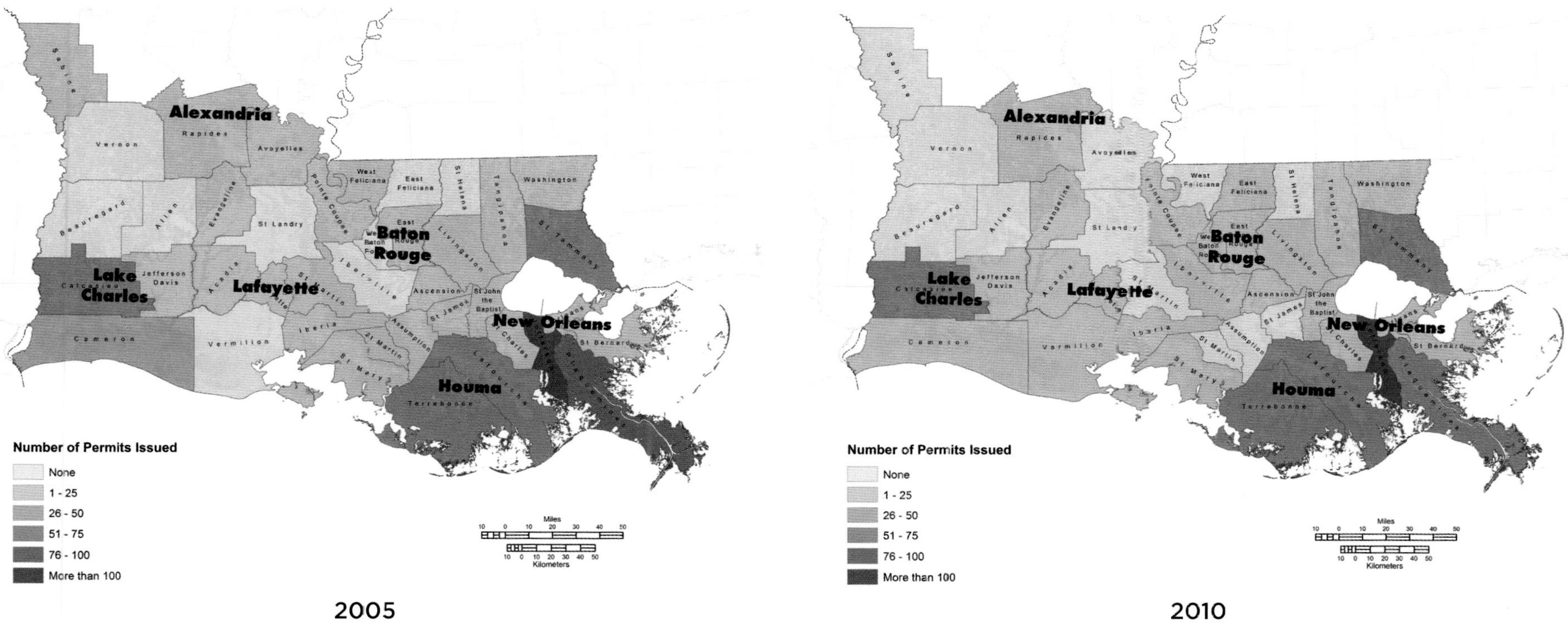

Fig. 74. Statewide Comprehensive Planning to Integrate Coastal Protection and Restoration Efforts

10 maps

Data Source: Coastal Protection and Restoration Authority of Louisiana

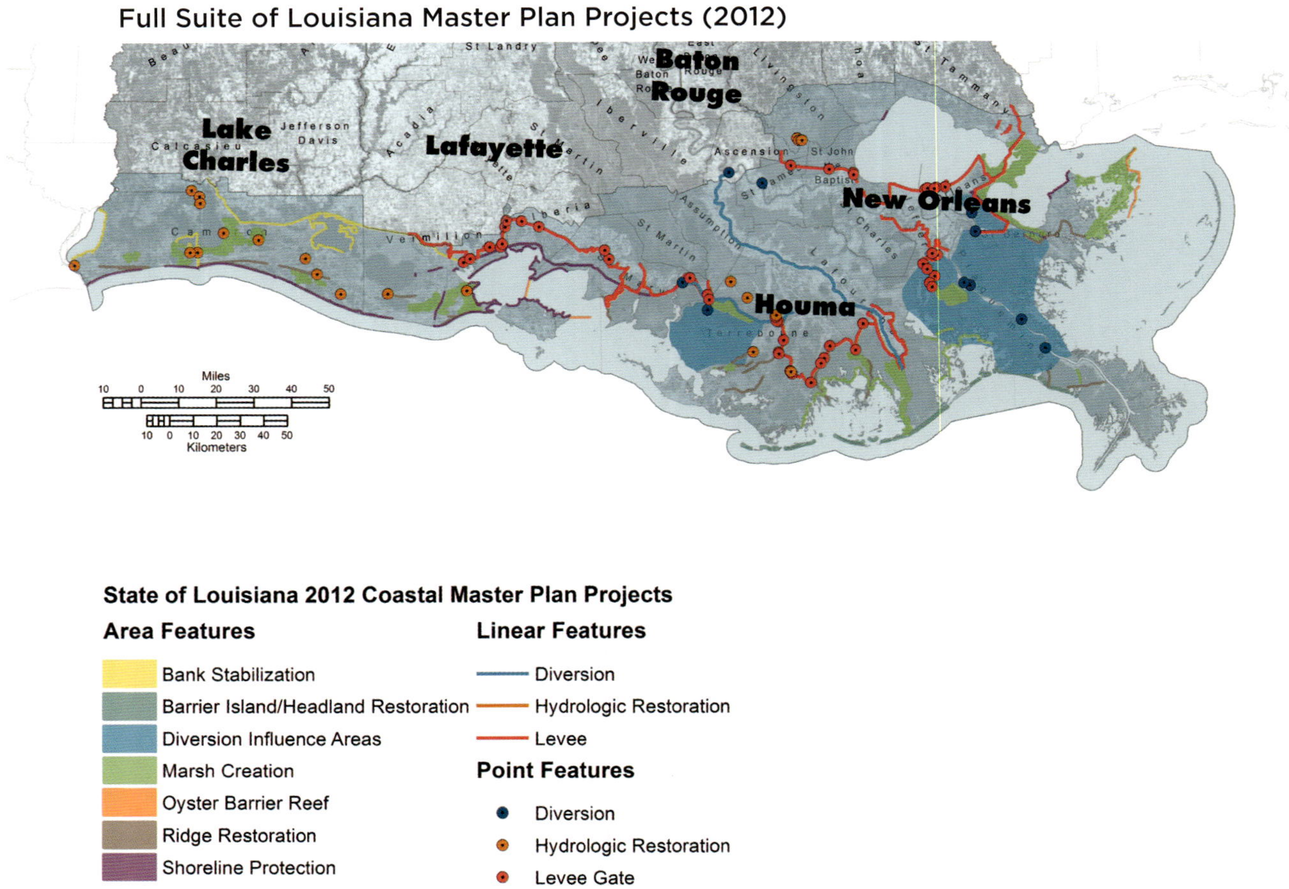

Ultimately, to reduce the rate of land loss and protect coastal communities from future tropical storm events, governmental and nongovernmental organizations must undertake a number of large-scale structural and nonstructural measures. The key to this effort must be a fully integrated coastal protection and restoration plan, one that addresses aspects of the physical, biological, and human environments. Louisiana's Comprehensive Master Plan for a Sustainable Coast was developed by the state in cooperation with an extensive network of state, federal, private, and academic researchers. This effort incorporates a variety of coastal restoration and protection projects in an attempt to maximize both protection of communities and restoration of the environment (fig. 74). Any projects funded by the state are required to show consistency with the goals and objectives of this plan.

Fig. 74. Statewide Comprehensive Planning to Integrate Coastal Protection and Restoration Efforts

Data Source: Coastal Protection and Restoration Authority of Louisiana

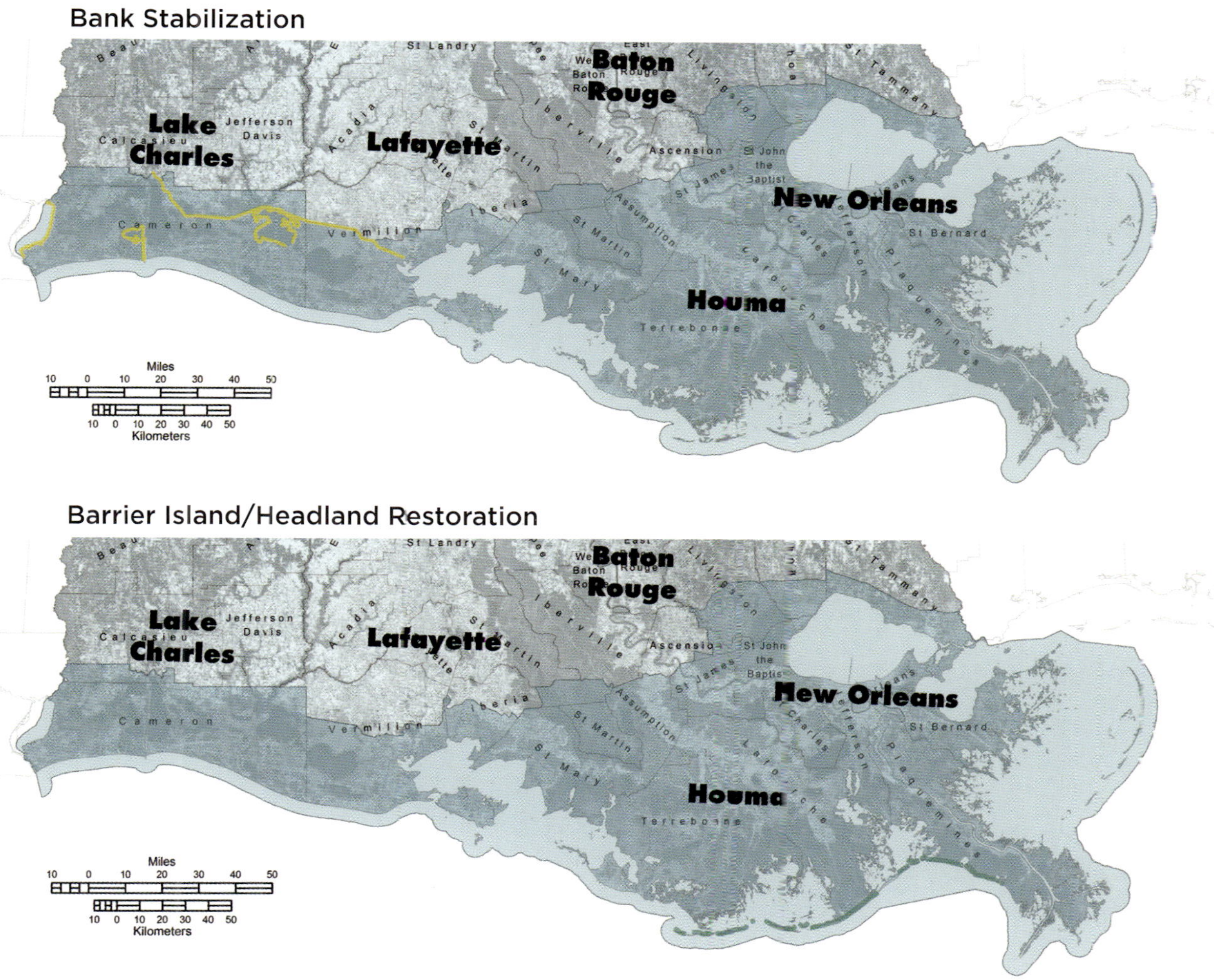

State of Louisiana 2012 Coastal Master Plan Projects

Area Features

- Bank Stabilization
- Barrier Island/Headland Restoration
- Diversion Influence Areas
- Marsh Creation
- Oyster Barrier Reef
- Ridge Restoration
- Shoreline Protection

Linear Features

- Diversion
- Hydrologic Restoration
- Levee

Point Features

- Diversion
- Hydrologic Restoration
- Levee Gate

Fig. 74. Statewide Comprehensive Planning to Integrate Coastal Protection and Restoration Efforts

Data Source: Coastal Protection and Restoration Authority of Louisiana

Diversions

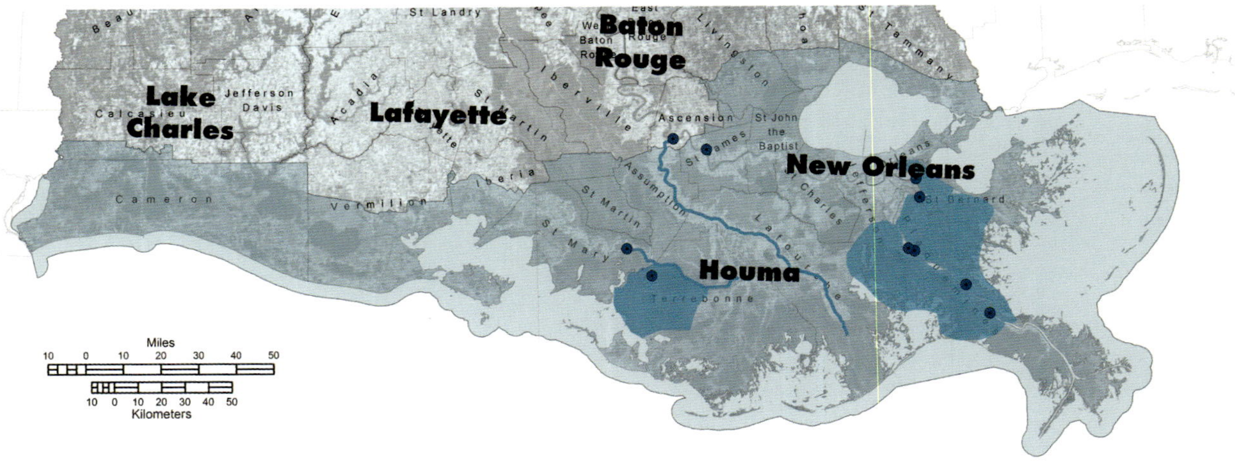

Hydro-Restoration

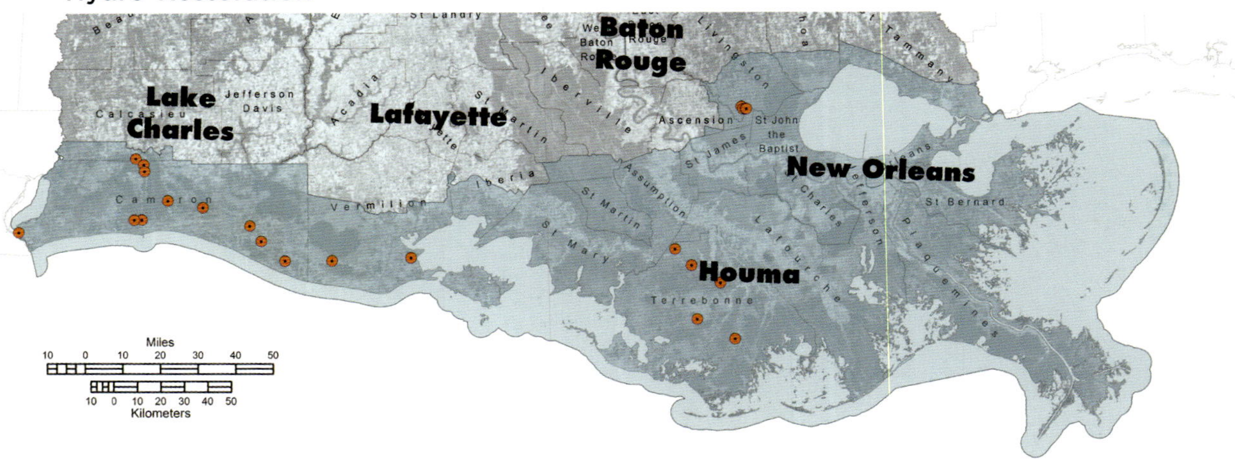

Levees

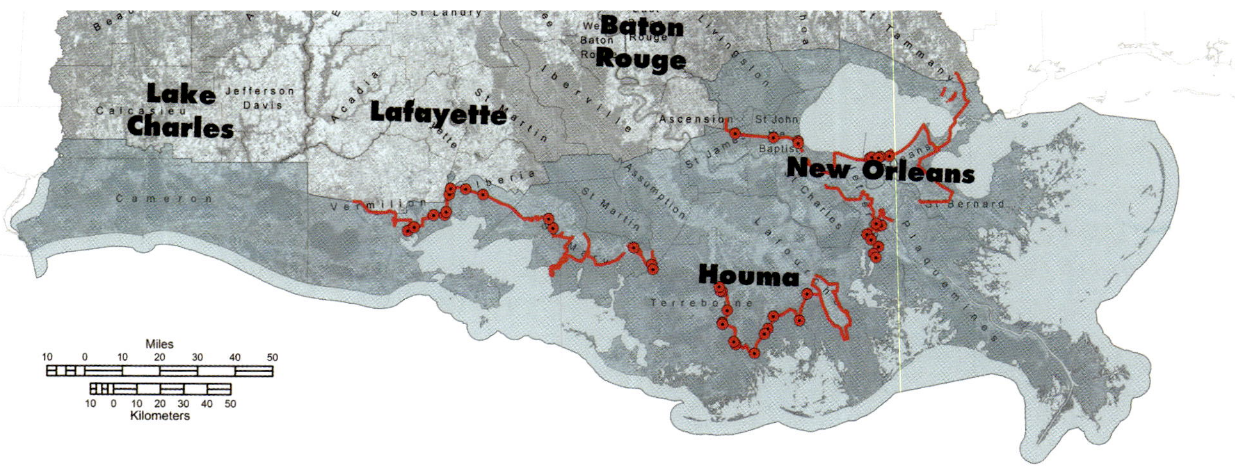

Fig. 74. Statewide Comprehensive Planning to Integrate Coastal Protection and Restoration Efforts

Data Source: Coastal Protection and Restoration Authority of Louisiana

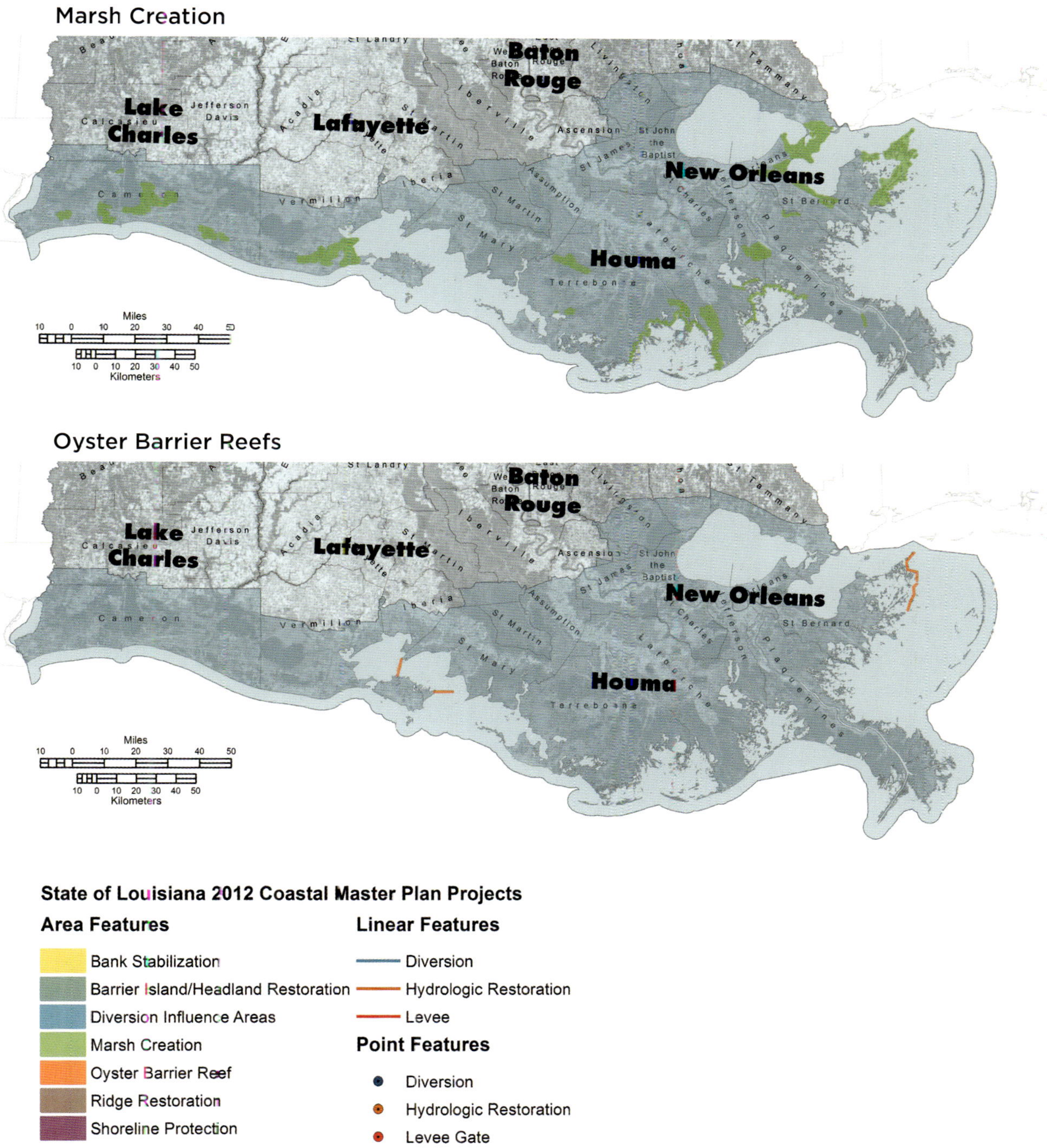

Fig. 74. Statewide Comprehensive Planning to Integrate Coastal Protection and Restoration Efforts

Data Source: Coastal Protection and Restoration Authority of Louisiana

Ridge Restoration

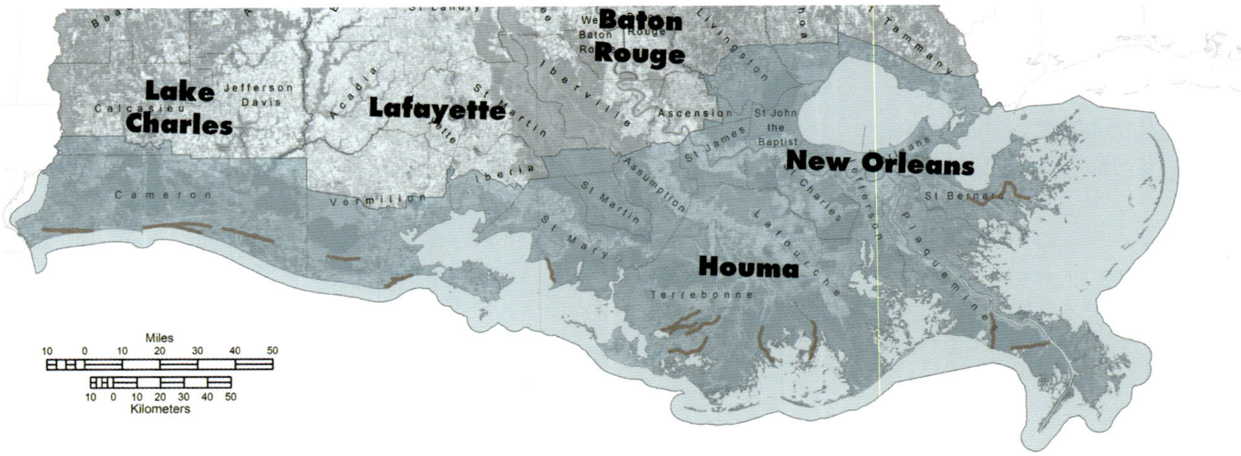

Shoreline Protection

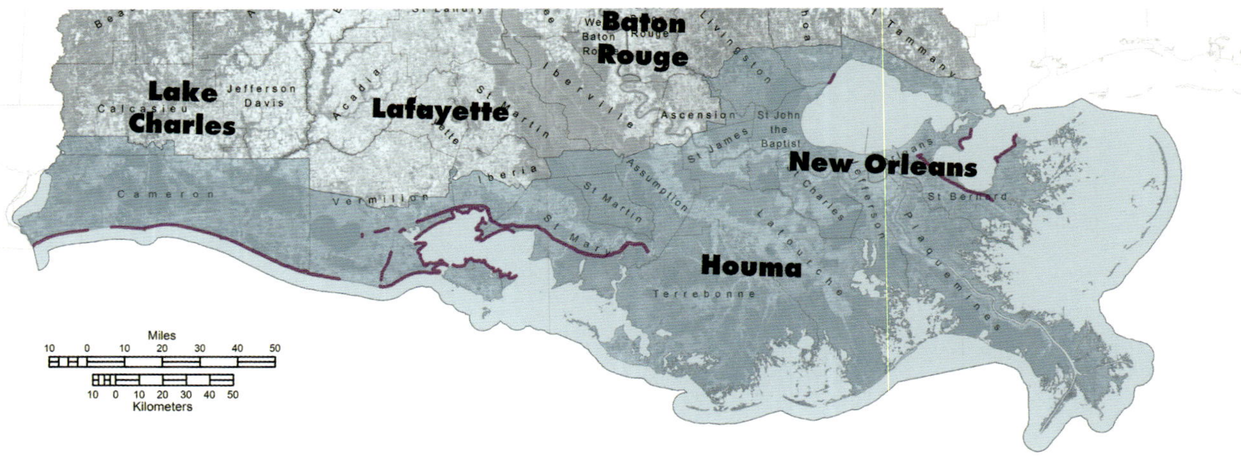

State of Louisiana 2012 Coastal Master Plan Projects

Area Features

- Bank Stabilization
- Barrier Island/Headland Restoration
- Diversion Influence Areas
- Marsh Creation
- Oyster Barrier Reef
- Ridge Restoration
- Shoreline Protection

Linear Features

- Diversion
- Hydrologic Restoration
- Levee

Point Features

- Diversion
- Hydrologic Restoration
- Levee Gate

Fig. 75. Coastal Homes Elevated to Base Flood Elevation or Higher

Data Source: State of Louisiana Office of Community Development

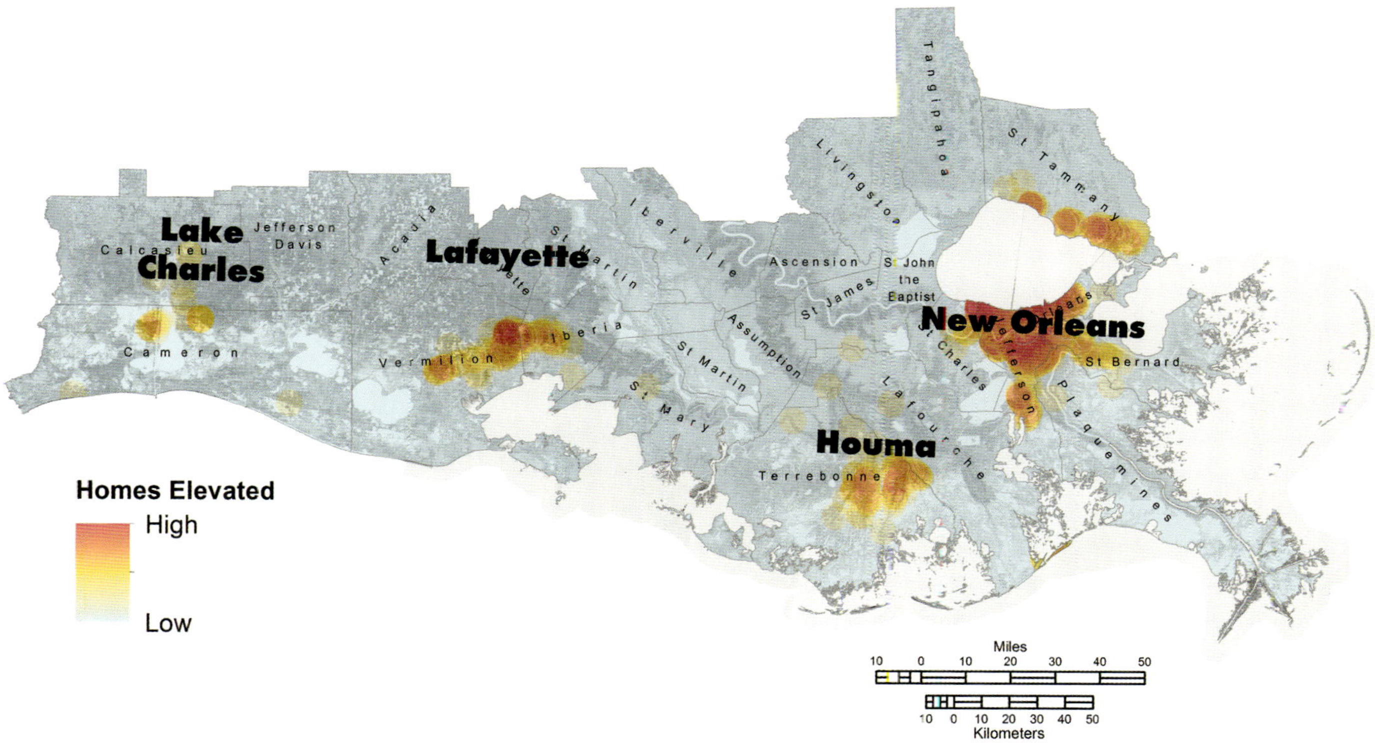

This comprehensive plan acknowledges that structural projects alone will not protect all coastal residents. Any future adaptations to life in the coastal zone must include an array of nonstructural protection features. At a minimum, such a plan should include options such as flood proofing and elevating homes, incorporating community resilience features into zoning and land-use plans, and developing a strong element of ecosystem-based protection in coastal restoration work. Each of these measures has been implemented to some degree across the coast. Following Hurricanes Katrina and Rita, federal grant monies were used to assist coastal residents in elevating their homes (fig. 75). The State of Louisiana recommends elevating residential structures to one foot above base flood elevation, which is equal to the 100-year flood depth.

Fig. 76. Coastal Homes Retrofitted with Individualized Mitigation Measures

Data Source: State of Louisiana Office of Community Development

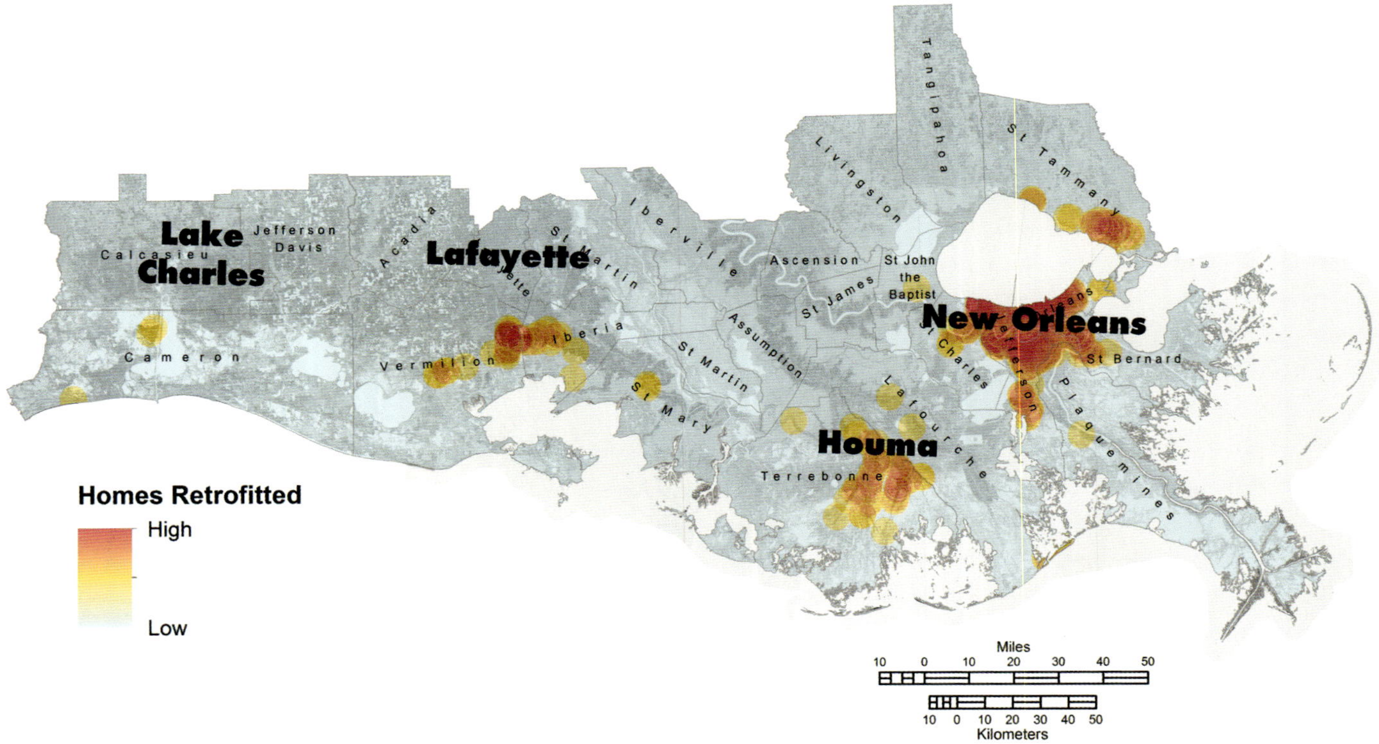

Other homes across the coast were flood proofed and retrofitted with individualized mitigation measures designed to protect them against other storm impacts, such as wind damage (fig. 76). Such mitigation measures include the installation of storm shutters, roof tie-downs, elevated air conditioning and heating units, strengthened doors, and anchored fuel tanks.

Fig. 77. Comprehensive Resiliency Pilot Programs

Data Sources: Louisiana Resiliency Assistance Program; State of Louisiana Office of Community Development

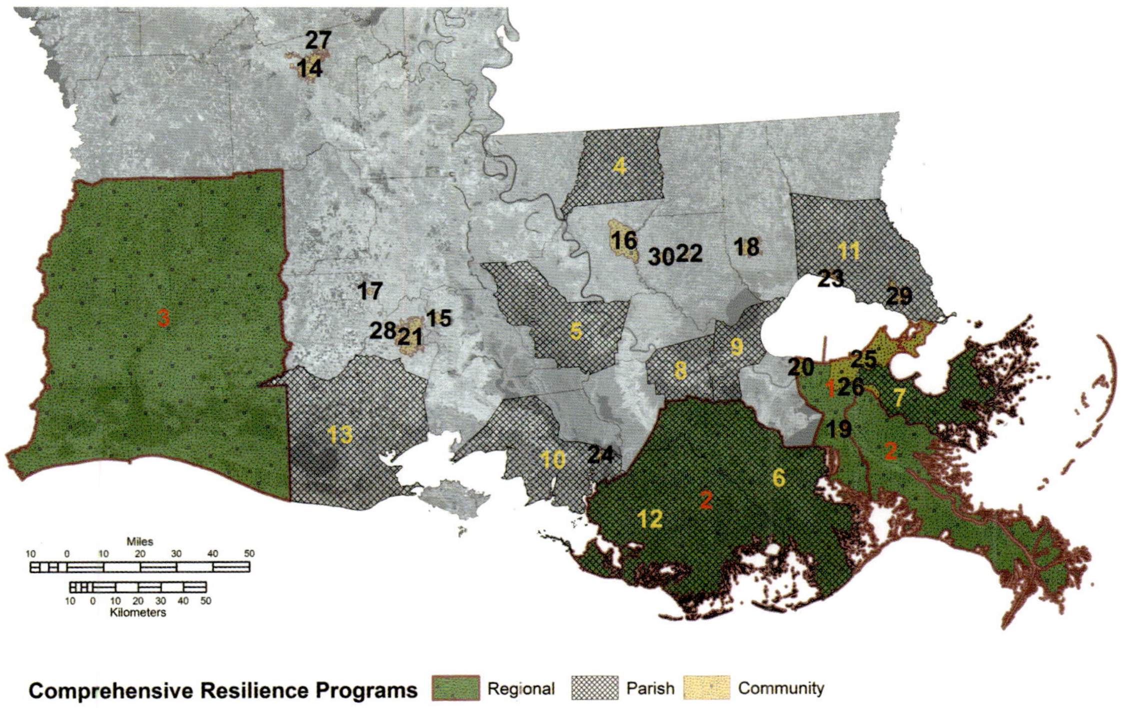

Regional Resilience Programs

1 Greater New Orleans Region-Urban Water Plan
2 Southeast Louisiana-Louisiana Wetland Action Program By Global Green
3 Southwest Louisiana-Southwest Louisiana Regional Housing Study

Parish Resilience Programs

4 East Feliciana Parish-Comprehensive Resiliency Plan
5 Iberville Parish-Development Code Revision, Zoning Ordinance And Zoning Map
6 Lafourche Parish-Comprehensive Resiliency Plan
7 St. Bernard Parish-Integrated Water Resource Management Plan
8 St. James Parish-Comprehensive Plan 2031
9 St. John The Baptist Parish-Development Management Program
10 St. Mary Parish-Unified Development Code
11 St. Tammany Parish-Comprehensive Resiliency Program
12 Terrebonne Parish-Vision 2030
13 Vermilion Parish-Comprehensive Resiliency Plan

Local Resilience Programs

14 Alexandria-THINKAlex Resiliency Plan
15 Breaux Bridge-Comprehensive Long-Range Resiliency Plan
16 Central-Resiliency Plan
17 Church Point-Long-Range Resiliency Plan
18 Hammond-Unified Development Code
19 Jean Lafitte-Jean Lafitte Tomorrow – Town Resiliency Plan
20 Kenner-Comprehensive Resiliency Plan
21 Lafayette-Community Watershed Development Plan
22 Livingston-Subdivision Regulations
23 Mandeville-Old Mandeville Town Center Plan
24 Morgan City-Zoning And Land Development Code And Land Use Development Plan
25 New Orleans-Main Street Resiliency Plan
26 New Orleans-Port Of New Orleans Design Resiliency Guidelines Manual
27 Pineville-Unified Development Ordinance And Zoning Map Revisions
28 Scott -2032 Plan
29 Slidell-Amendment To Code Of Ordinances
30 Walker-Blueprint Of A City: A Community Vision For Walker, La

Effective planning is necessary for communities and regions to prepare for, and recover from, extreme hazard events. To this end, many communities, parishes, and regional organizations have developed pilot resilience programs. These programs are designed to build community resilience through such strategies as updated comprehensive plans, zoning ordinances, and building code enforcement (fig. 77).

Fig. 78. Ecosystem-Based Adaptation Measures

Data Sources: Coalition to Restore Coastal Louisiana; Coastal Protection and Restoration Authority of Louisiana; NOAA Restoration Center

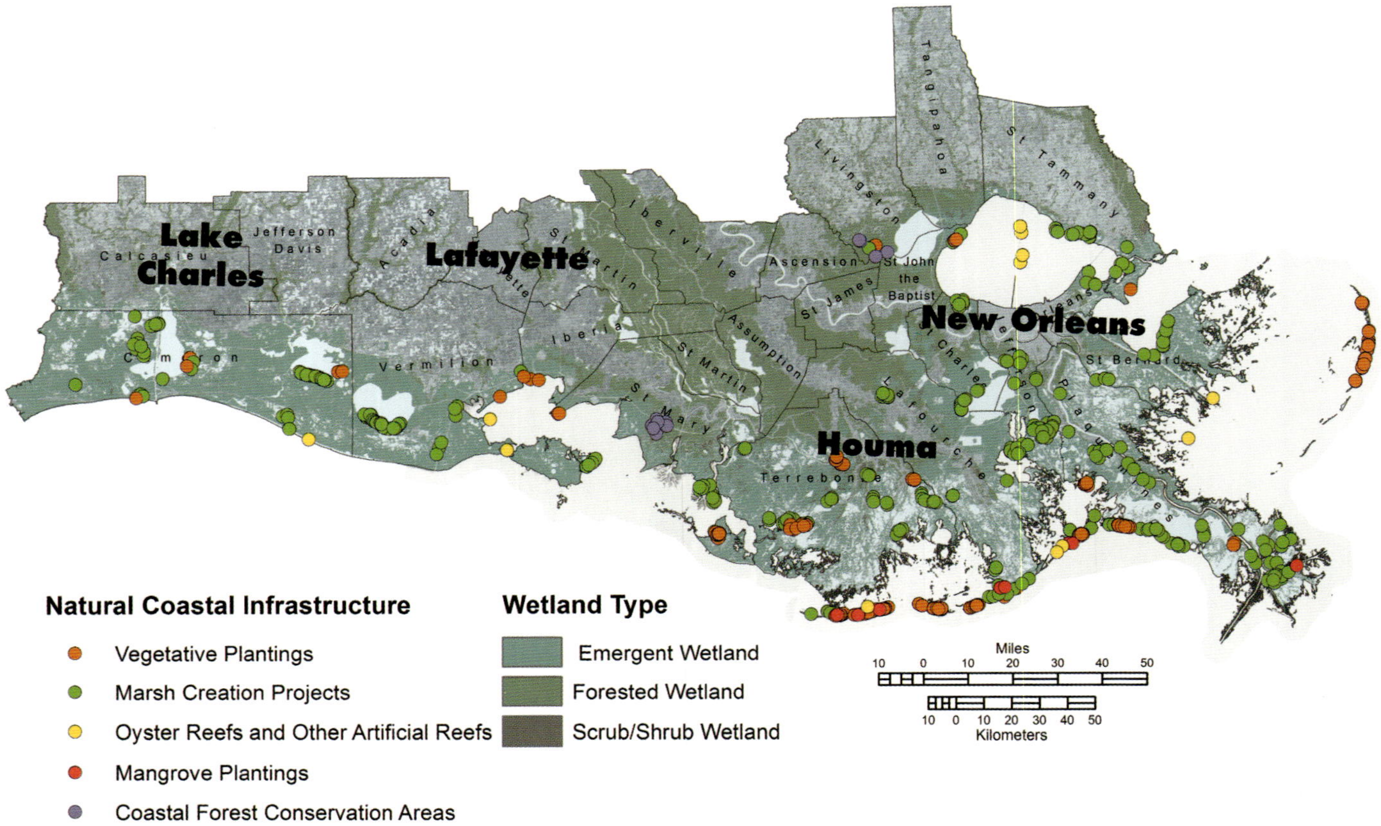

Louisianans have had a long history of coexisting with the environment. They have relied on it for physical and spiritual sustenance. They have relied on it for employment. And they have relied on it for protection. Coastal communities have persisted not despite the ever-changing environment, but because of it. If these communities are to continue to thrive in the coastal zone, they must use the environment to provide the same valuable protection that it has provided historically. To this end, ecosystem-based adaptation measures must be a key element of any coastal protection and restoration plan (fig. 78). The wetlands of coastal Louisiana are the first and best defense against the tropical storms that constantly threaten the state. Protecting what remains of the state's coastal forests, rebuilding the marshes, and replanting vegetation to hold the ground in place are all projects that are going on today. Innovative strategies such as the construction of oyster reefs are being developed. Mangroves are being planted to provide an additional line of defense against tropical storms. Both governmental and nongovernmental organizations are leading these efforts. Both public workers and private citizens are doing this work. It is efforts such as these that will eventually determine how resilient coastal Louisiana is and can be.

CONCLUSION

Coastal Louisiana is and always has been a place of constant change. Places are products of time and space experienced by people through the lenses of both history and geography. When viewed at the scale of human perception, certain locales are regarded as being more stable than others, seemingly unchanging. Occasional shocks may come quickly and powerfully but communities often re-stabilize just as quickly, with the structure and infrastructure that sustain that community becoming fully intact. In such places, landscape-level environmental change generally occurs slowly. But even in such seemingly stable geographical settings, small fluctuations to the underlying structures of society are always taking place. Within every community, economic and social changes, sometimes driven by larger national or global influences, sometimes by more subtle changes in local mores and values, are constantly changing the population in subtle ways. Slow-moving crises such as climate change may also directly and indirectly force change in local communities, even if this change is only experienced at a multigenerational scale.

Areas where the environment is perceived as a constant, a stable base on which economic and societal change plays out, can be conceptualized as one extreme on the scale of social and environmental change. Coastal areas would most certainly reside at the other end of the spectrum. In Louisiana's coastal zone, societal and economic change occurs in the midst of tremendous environmental variability. Singular events such as hurricanes and river floods are a persistent threat to coastal communities as is the threat of drought to communities heavily reliant upon agriculture. Slow burn, creeping environmental change, in the form of sea-level rise, subsidence, and coastal land loss represents a lesser known, yet constant threat to the well-being of all coastal communities, even those seemingly well protected within the coastal levee systems that encase the vast majority of residents. Such threats occur beyond the time frame of annual fiscal budgets and election cycles. This creates a situation where the threats seem somewhat more elusive or vague, as do the means of responding to them. Because communities view and experience these events through the lens of time, they are more difficult to adapt to. Similarly, the success or failure of adaptive measures is difficult to measure in the short term, if at all.

Despite the difficulties involved, communities still must constantly prepare for, react to, and, ultimately, adapt to these threats. In some cases, communities adapt in ways that enhance their resilience, allowing them to persist, even thrive, in such a perilous location. In other cases, communities may lose their resilience, changing in ways that are not sustainable in the long term. In every case, the intensity and duration of environmental change most certainly act as powerful drivers of societal change. The interaction of various change agents creates new social trajectories. These interactions may reinforce current trends, or they may change the existing inertia of change, establishing what can best be seen as a new normal.

Without effective adaptation, storm events often just accentuate preexisting trends, for better or for worse. Hurricane Katrina, the storm with the highest visibility in the public consciousness, had a devastating impact on the population of several southeast Louisiana parishes, particularly Orleans, Plaquemines, and St. Bernard. While the effects of the storm on population levels cannot be overstated, this region was already in the midst of a radical population reorganization occurring over the several decades prior to the storm. A national trend of suburbanization combined with social factors such as school desegregation drove many white residents out of New Orleans and into neighboring parishes and those on the northern shore of Lake Pontchartrain starting in the 1960s. By the 1990s, population began to decline in some of these neighboring parishes, such as St. Bernard, even as the Northshore communities continued to expand.

These historical trends would suggest that the population of southeast Louisiana was primed for collapse, and when Hurricane Katrina devastated the area, much of the population did, in fact, flee, a great many with no intentions of returning. Many of those that did return came back with lower degrees of inherent resilience. Those pre-existing conditions that drove population change prior to the storm still existed and in some cases worsened. Crime rates, for example, a long-time problem in the New Orleans metro area, increased further following the storm, driving ever more people to suburban Northshore communities and beyond. In fact, the greatest population growth in the region in the years following the storm occurred not in the immediately adjacent Northshore parishes, but rather in the suburban parishes to the south and east of Baton Rouge. Here, antecedent suburbanization from the Baton Rouge urban core combined with an influx of New Orleans residents intent on resettling farther afield from the circum-Pontchartrain region, where the threat of tropical storm events and coastal flooding is significantly lessened.

While Katrina could certainly be considered an extreme outlier event, due to the magnitude of the storm, the immediate impacts on life and property, and the slow rate of recovery, this same trend holds true for many other large tropical storm events. Hurricane Andrew, for example, cut a swath through central Louisiana in the late summer of 1992, making landfall in St. Mary Parish, an area already reeling from a devastating economic downturn in the oil industry occurring in the previous decade. By the time of the 1990 census, the population of St. Mary Parish had already decreased significantly from 1980 levels. By 2000, that trend was exacerbated by the effects of the storm. The population loss that had begun with the loss of industry employment was compounded by the impacts of Hurricane Andrew.

These examples highlight the role of crises in maintaining or eroding community resilience. The effects of these shocks compound to the point where a community may fail to rebound. A community that has withstood numerous large exogenous shocks through the years may collapse when faced with a single small endogenous shock. Similarly, a community that has coped with a long-term, slow exogenous shock, such as coastal land loss, by not adapting to changing conditions may begin to lose its inherent resilience in a slow but inevitable way. This community may be impacted by several large shocks from which it physically recovers, but without adaptation, that community is continually being eroded along with the ground on which it was built.

The degree to which a society is able to effectively cope with and adapt to change is clearly a crucial factor in determining how resilient that society is. While social changes and environmental threats present ongoing challenges for coastal populations and the communities that sustain them, loss of community resilience is not an inescapable outcome. In other words, despite the ever-increasing threats posed by external forces, both social and environmental, people and society are not fixed on a path of decline and loss. Path dependence is real but the endpoint is not inevitable.

In biological terms, adaptation is defined as the development of physical and behavioral characteristics that allow organisms to survive and persist in their habitats. In terms of the human environment, this definition functions at multiple scales, from the individual to the community up to and including the broader society. The ultimate key to the sustainability of human development in Louisiana's coastal zone is the ability of communities to not just react, but to cultivate characteristics that can effect positive change in the physical and social environment. Positive change is largely driven by the ability of communities to effectively adapt to changing conditions in new and innovative ways.

Extreme events and environmental shocks are often thought of as a cause of reduced resilience in coastal communities. Hurricanes and river floods, for example, may have an immediate, devastating impact on the population, economy, and built environment of a region. To compound matters, when viewed over long timescales, these events begin to take on characteristics commonly associated with slow-burn events. For example, periodic events such as river floods that occur on a regular, though erratic, schedule begin to create high levels of chronic stress that, without mitigation, corrode regional resilience.

The reality of the situation is not so cut and dry, however. It is the response to and recovery from hazard events that determines how resilient a community ultimately becomes. While large storm events can potentially drive down community resilience through a variety of mechanisms, they also present an opportunity for change. One shock often brings the region together. Multiple shocks of a similar kind can enhance awareness of problems and lead to systemic changes that make regions more disaster-proof and thereby more resilient. Attempts to both recover from and inoculate against these threats may galvanize communities and their leaders, creating a sense of regional unity. The question then becomes, is it possible to leverage this shared purpose to allow communities to adapt in such a way so as to chart a new sustainable path forward?

It is important to recognize that change is a constant, unavoidable reality of life in Louisiana's coastal zone. Community resilience is a process that is built over time as individuals and groups adapt and react to this change. Community resilience is often mistakenly conceptualized as a static, inherent quality of a community. Researchers derive quantitative measures that assign a specific value to each geographical unit within a study area, allowing residents and researchers alike to state that this community is more resilient than that community. This analytical framework has many benefits, chief among these being the ready availability of secondary data that can be used and manipulated and visualized in such a way as to identify local and regional patterns in social vulnerability and community resilience. Indeed, improvements in geospatial technology and data-capturing techniques have created an environment where tremendous amounts of data can be used to derive any number of customized local resilience indices which capture in some way those social and economic factors that make one community more likely to recover from an extreme hazard event than another community. Such values are often thought of as the baseline status of a community, the initial conditions of some social model being played out in real time within the communities embedded in the landscape. When divorced from historical reality, these conditions are often treated as the equilibrium condition against which any future changes can be measured.

Any appearance of equilibrium in a social system is ultimately illusory. Social systems that seem to be stable are merely exhibiting a short-term collective capacity to effectively adapt to ever-changing social and environmental conditions. When examined over a larger timescale, over a period of time ranging from decades to generations, these moments of relative stability can be seen as singular, ephemeral points in time that are a product of multiple historical decisions and adjustments to changing circumstances. Some of these adjustments may drive society to higher levels of stability while others may weaken society and reduce levels of societal resilience. Understanding the historical changes that have shaped present-day communities is vital to any planning effort whose goal or purpose is to protect and preserve these communities and the people who reside in them.

The question facing coastal communities is not only how to survive in a place of constant change, but how to thrive. This is more than a question of assuring that the lifelines that sustain society are maintained. While the role of protecting critical infrastructure and essential facilities cannot be overstated, it must be remembered that community is much more than the sum of the individuals and structures that exist in any one place. When we speak of preserving community, we are, at some level, speaking of preserving historical continuity. It is this continuity that ties the people to the landscape. If we lose that sense of continuity, the idea that a community is a place that is tied to an underlying cultural uniqueness, then community resilience is lessened. Viewed in this light, resilience is about maintaining and enhancing the past in such a way as to point towards a new, stronger future.

The cultural heritage of coastal Louisiana is inextricably tied to the landscape and ecology of the coast. This relationship may shift through time. Trappers who once worked the wetlands for valuable mammal pelts now hunt the invasive nutria that have caused irreparable harm to the ecosystem. Reality television has altered, or perhaps reinforced, the public notion of what it means to reside in the swamps and marshes of coastal Louisiana. As a result, ecotourism opportunities are on the rise. Agritourism has the potential to transform local farm communities. Commercial fishermen lead excursions into the bays and bayous of the region, bringing their traditional ecological knowledge to the broader public. Yet, these cultural adaptations will be rendered meaningless should the environment they depend on fragment, degrade, or vanish into the waters of the Gulf of Mexico. This is the challenge of the twenty-first century for residents of coastal Louisiana and coastal scientists and planners working to preserve this valuable social and ecological system.

While change is inevitable, society's reactions are not. Human agency is a powerful tool that allows societies to transcend their history and plot a new course into the future. Historical momentum, while often difficult to overcome, can be shifted in new and powerful ways. Small, seemingly insignificant changes can alter the inertia of the social system, setting society on a new, more resilient path, one built on the knowledge of the past, the resources of the present, and the innovations of the future.

REFERENCES

Adams, D. C. 2013. And the ocean came up on land: perceptions of adaptive capacity of cattle ranching in Vermilion Parish, Louisiana. Master's thesis, University of Texas at Austin. Retrieved from http://repositories.lib.utexas.edu/handle/2152/28674

Becnel, T. A. 1962. A history of the Louisiana shrimp industry, 1867–1961. Master's thesis, Louisiana State University.

Campanella, Richard. 2010. *Delta Urbanism: New Orleans.* Chicago, IL: American Planning Association.

Chesney, E. J., D. M. Baltz, and R. G. Thomas. 2000. Louisiana estuarine and coastal fisheries and habitats: Perspectives from a fish's eye view. *Ecological Applications* 10(2): 350–66.

Coelho, K. 2006. The footprint of the offshore oil industry on community institutions. *Research in Economic Anthropology* 24: 123–62.

Colten, C. E. 2012. An incomplete solution: Oil and water in Louisiana. *Journal of American History* 99(1): 91–99.

———. 2014. *Scenario Building Workshops.* Baton Rouge, LA: Water Institute of the Gulf.

Colten, C. E., A. A. Grismore, and J. R. Z. Simms. 2015. Oil spills and community resilience: Uneven impacts and protection in historical perspective. *Geographical Review* 105(4): 391–407.

Colten, C. E., J. Hay, and A. Giancarlo. 2012. Community resilience and oil spills in coastal Louisiana. *Ecology and Society* 17(3): 5.

Colten, C. E., R. W. Kates, and S. B. Laska. 2008. *Community Resilience: Lessons from New Orleans and Hurricane Katrina.* CARRI Research Report 3, Oak Ridge National Laboratory.

Couvillion, B. R., G. D. Steyer, H. Wang, H. J. Beck, and J. M. Rybczyk. 2013. Forecasting the effects of coastal protection and restoration projects on wetland morphology in coastal Louisiana under multiple environmental uncertainty scenarios. *Journal of Coastal Research* 67(spl): 29–50.

Couvillion, B. R., J. A. Barras, G. D. Steyer, W. Sleavin, M. Fischer, H. Beck, N. Trahan, B. Griffin, and D. Heckman. 2011. *Land Area Change in Coastal Louisiana from 1932 to 2010.* U.S. Geological Survey.

CPEX (Center for Planning Excellence). 2012. *Best Practices Manual for Development in Coastal Louisiana.* Baton Rouge, LA: CPEX. Page 120.

CPRA (Coastal Protection and Restoration Authority). 2012. *Louisiana's Comprehensive Master Plan for a Sustainable Coast.* Baton Rouge, LA: CPRA. Page 188.

Crosset, K. M., T. J. Culliton, P. C. Wiley, and T. R. Goodspeed. 2004. *Population Trends along the Coastal United States, 1980–2008.* National Oceanic and Atmospheric Administration Report.

Dalbom, C., S. A. Hemmerling, and J. A. Lewis. 2014. *Community Resettlement Prospects in Southeast Louisiana: A Multidisciplinary Exploration of Legal, Cultural, and Demographic Aspects of Moving Individuals and Communities* (issue paper). New Orleans, LA: Tulane Institute on Water Resources Law & Policy. Page 49.

Davis, Donald. W. 2010. *Washed Away? The Invisible Peoples of Louisiana's Wetlands.* Lafayette, LA: University of Louisiana at Lafayette Press.

Dolan, A. H., and I. J. Walker. 2006. Understanding vulnerability of coastal communities to climate change related risks. *Journal of Coastal Research* 39 (spl): 1316–1323.

Eisterhold, J. A. 1972. Lumber and trade in the Lower Mississippi Valley and New Orleans, 1800-1860. *Louisiana History* 13(1): 71-91.
Estaville, L. 1993. The Louisiana-French homeland. *Journal of Cultural Geography* 13(2): 31-45.
Germann, J. J. 1990. *Louisiana Post Offices*. Lake Grove, OR: The Depot.
Keithy, W. 1991. *Louisiana Seafood Industry Study: A Summary*. New Orleans, LA: Louisiana Seafood Promotion and Marketing Board.
Laska, S., G. Wooddell, R. Hagelman, R. Gramling, and M. T. Farris. 2005. At risk: the human, community and infrastructure resources of coastal Louisiana. *Journal of Coastal Research* 44 (spl): 90-111.
Lenton, T. M. 2013. What early warning systems are there for environmental shocks? *Environmental Science & Policy* 27: S60-S75.
Levine, J. N., A.-M. Esnard, and A. Sapat. 2007. Population displacement and housing dilemmas due to catastrophic disasters. *Journal of Planning Literature* 22(1): 3-15.
Lopez, J. A. 2009. The multiple lines of defense strategy to sustain coastal Louisiana. *Journal of Coastal Research* 54(spl): 186-97.
Mancil, E. 1969. Some historical and geographical notes on the cypress lumbering industry in Louisiana. *Louisiana Studies* 8(1): 14-25.
——. 1972. Historical geography of industrial cypress lumbering in Louisiana. Ph.D. dissertation, Louisiana State University.
Marino, E. 2012. The long history of environmental migration: Assessing vulnerability construction and obstacles to successful relocation in Shishmaref, Alaska. *Global Environmental Change* 22(2): 374-81.
Marks, B. 2012. The political economy of household commodity production in the Louisiana shrimp industry. *Journal of Agrarian Change* 12 (2-3): 227-51.
McGuire, T. 2006. Oil and gas in south Louisiana. *Research in Economic Anthropology* 24: 63-87.
Mileti, D. S., and E. Passerini. 1996. A social explanation of urban relocation after earthquakes. *International Journal of Mass Emergencies and Disasters* 14(1): 97-110.
Millennium Ecosystem Assessment (Program). 2005. *Ecosystems and Human Well-Being: Synthesis*. Washington, D.C.: Island Press.
Moore, J. M. 1983a. The cypress lumber industry of the Lower Mississippi Valley during the colonial period. *Louisiana History* 24(1): 25-47.
——. 1983b. The cypress lumber industry of the Old Southwest and public land law, 1803-1850. *Journal of Southern History* 49(2): 203-22.
Noble, I., S. Huq, Y. Anokhin, J. Carmin, D. Goudou, F. Lansigan, B. Osman-Elasha, A. Villamizar, J. Ayers, F. Berkhout, K. Dow, H. Fussel, J. Smith, K. Tierney, and H. Wright. 2014. Adaptation needs and options. Chap. 14 of *Working Group II of the Intergovernmental Panel on Climate Change Fifth Assessment Report*.
Norgress, R. E. 1947. The history of the cypress lumber industry in Louisiana. *Louisiana Historical Quarterly* 30(3): 3-83.
Osti, R., S. Tanaka, and T. Tokioka. 2009. The importance of mangrove forest in tsunami disaster mitigation. *Disasters* 33(2): 203-13.
Padgett, H. R. 1969. Physical and cultural associations on the Louisiana coast. *Annals of the Association of American Geographers* 59(3): 481-93.
Pallagst, Karina, Thorsten Wiechmann, and Cristina Martinez-Fernandez, eds. 2013. *Shrinking Cities: International Perspectives and Policy Implications*. Routledge.
Paton, D. 2006. Disaster resilience: Building capacity to co-exist with natural hazards and their consequences. In Douglas Paton and David Johnston, eds., *Disaster Resilience: An Integrated Approach*. Springfield, IL: Charles C. Thomas Pub. Pages 3-10.
Pendall, R., K. A. Foster, and M. Cowell. 2007. *Resilience and Regions: Building Understanding of the Metaphor*. Berkeley, CA: Institute of Urban and Regional Development.
Peyronnin, N., M. Green, C. P. Richards, A. Owens, D. Reed, J.

Chamberlain, D. G. Groves, W. K. Rhinehart, and K. Belhadjali. 2013. Louisiana's 2012 coastal master plan: Overview of a science-based and publicly informed decision-making process. *Journal of Coastal Research* 67(spl): 1-15.
Pulsipher, A. 2006. *Accounting for Socioeconomic Change from Offshore Oil and Gas: Cumulative Effects on Louisiana's Coastal Parishes, 1969-2000.* Baton Rouge, LA: U.S. Dept. of the Interior, Minerals Management Service, OCS Study, MMS 2006-030.
Rehder, John B. 1999. *Delta Sugar: Louisiana's Vanishing Plantation Landscape.* Baltimore: Johns Hopkins University Press.
Schetter, C. D., and C. Dolbier. 2011. Resilience in the context of chronic stress and health in adults. *Social and Personality Psychology Compass* 5(9): 634-52.
Smithers, J., and B. Smit. 1997. Human adaptation to climatic variability and change. *Global Environmental Change* 7(2): 129-46.
Thatcher, C. A., J. C. Brock, and E. A. Pendleton. 2013. Economic vulnerability to sea-level rise along the northern U.S. Gulf Coast. *Journal of Coastal Research* 63(spl): 234-43.
Theriot, Jason P. 2014. *American Energy, Imperiled Coast: Oil and Gas Development in Louisiana's Wetlands.* Baton Rouge, LA: Louisiana State University Press.
Titus, J. G., K. E. Anderson, et al. 2009. *Coastal Sensitivity to Sea-Level Rise: A Focus on the Mid-Atlantic Region.* U.S. EPA: National Service Center for Environmental Publications.
Turner, B. L., R. E. Kasperson, P. A. Matson, J. J. McCarthy, R. W. Corell, L. Christensen, N. Eckley, J. X. Kasperson, A. Luers, M. L. Martello, C. Polsky, A. Pulsipher, and A. Schiller. 2003. A framework for vulnerability analysis in sustainability science. *Proceedings of the National Academy of Sciences* 100(14): 8074-8079.
Usner, D. H. 1979. From African captivity to American slavery: The introduction of black laborers to colonial Louisiana. *Louisiana History* 20 (1): 25-48.
Vaughan, E. 2003. Louisiana sugar: A geohistorical perspective. Ph.D. dissertation, Louisiana State University.

INDEX